T0284005

Charge

Frank Close FRS is an eminent research theoretical physicist in nuclear and particle physics. Currently Emeritus Professor of Physics at Oxford University and a Fellow of Exeter College, he was formerly the Head of the Theoretical Physics Division at the Rutherford Appleton Laboratory. He served as Chair of the UK Space Exploration Working Group 2007 which culminated with Tim Peake's launch to the ISS. He is the author of several books, including the best-selling *Lucifer's Legacy* (2000), and his highly acclaimed biography of the Higgs Boson *Elusive* (2022). His other books include *Antimatter* (2018), *Neutrino* (2011), *Eclipse: Journeys to the Dark Side of the Moon* (2017), and *A Very Short Introduction to Nuclear Physics* (2015), *Particle Physics* (2004), and *Nothing* (2009). In 2013, Professor Close was awarded the Royal Society Michael Faraday Prize for communicating science, and was elected Fellow of the Royal Society in 2021.

Charge

Why does gravity rule?

FRANK CLOSE

OXFORD UNIVERSITY PRESS

Great Clarendon Street, Oxford, OX2 6DP,
United Kingdom

Oxford University Press is a department of the University of Oxford.
It furthers the University's objective of excellence in research, scholarship, and education by publishing worldwide. Oxford is a registered trade mark of Oxford University Press in the UK and in certain other countries

Published in the United States of America by Oxford University Press
198 Madison Avenue, New York, NY 10016, United States of America

British Library Cataloguing in Publication Data

Data available

Library of Congress Control Number: 2023948648

ISBN 9780198885054

DOI: 10.1093/oso/9780198885054.001.0001

Printed and bound in the UK by
Clays Ltd, Elcograf S.p.A.

Contents

1

A 2500-YEAR-OLD MYSTERY

If we could approach an atom from afar, our first sight would be a swarm of tiny electrically charged particles. These are electrons. The electron appears to be one of the basic letters in nature's alphabet; if there are more fundamental constituents lurking within an electron, we have yet to find any hint of them. Each electron is identical in shape, size, and mass, and carries the same tiny amount of negative electric charge.

While the units of mass, length, and time are familiar—grams and kilograms, metres and kilometres, seconds and years—the measure of electric charge is less so. The unit of charge is a 'coulomb', often written as a capital C.

The amount of charge on a single electron is trifling. It would take more than six trillion of them to make one micro-coulomb—a millionth of a coulomb—about the amount of charge that you could feel as a slight static shock. However, in even a mote of dust there are more atoms than there are stars in our galaxy, and each of those atoms contains several electrons. Although a single electron's charge is insignificant, the result of a small quantity being multiplied innumerable times can be startling.

For example, with each breath, you inhale about ten billion trillion atoms of oxygen, each containing several negatively charged electrons. In all, this amounts to about 15,000 coulombs of electric charge, enough to spark 1000 bolts of lightning. Yet you are unaware of it. Your hairs don't suddenly stand on end; you don't emit sparks, nor do you experience an electric shock. And some shock! If you are breathing steadily the intake corresponds to an electric current of some 3000 amps, enough to kill you and save your executors the cost of cremation. Yet here we are. Nothing reveals this remarkable electricity deep within us. The reason why remains a mystery.[1]

Theories attempting to explain it abound, and partial answers have been found. At root, however, this remains one of the most basic unanswered questions in fundamental science.

****

Electrons might appear to be fanciful particles lurking within atoms and otherwise irrelevant, but they are like a gateway to the modern world. Electrons are the most fundamental easily accessible carriers of electric charge. When you charge the battery in your laptop, phone, or even your electric car, you're storing electrons for later use.

Electric current is the flow of electrons. This flow could be through computer chips, along the strands of your central nervous system transmitting instructions from brain to limbs,

[1] The derivation of these numbers is given in the appendix: 'An electric shock'.

through the overhead wires or third rails that power electric trains, or amidst any of the myriad applications of our modern electrical industries. Much of modern technology is basically driven by electrons on the move. A current of one ampere, or *amp*, is defined to be the flow of one coulomb of charge in a second. That corresponds to 6.24 billion billion electrons passing a single point in the time of a heartbeat.

These enormous numbers of electrons in motion can make even small amounts of current lethal. It is traditional to focus on milliamps—one-thousandth of an amp—when assessing the effect of electric current on the human body. Just 10 milliamps can be enough to give you a painful shock, whereas currents between 100 and 200 milliamps are lethal. If you touch a cable carrying more than 10 milliamps of current, your muscles contract so powerfully that you will be unable to let go of the wire that is shocking you: the muscles controlling the lungs are affected and soon you stop breathing. At around 100 milliamps the muscles controlling the heart are affected; its ventricles take on an uncontrolled twitching that leads to death. Above 200 milliamps survival chances are good, however, as the heart's muscles become clamped, which prevents the onset of ventricular fibrillation. However, severe burning might result.

The physiology of electric current is a complex subject, but the above dramas are sufficient to highlight our paradox. A current of 100 milliamps is enough to cause death, while larger currents fry tissues and organs, yet inhaling 15,000 coulombs in a single breath corresponds to receiving a current of 3000 amps for several seconds, and we do this all the time. A single thunderbolt is enough to kill, and you have in effect just absorbed the effects of a thousand of them. How do you survive? Why are there not sparks flying around your head giving you an appearance like some manic cartoon character as you are turned to charcoal?

The reason is that in every atom, at the centre of this swarm of negative charges, is a massive—several thousand times more so than an individual electron—dense lump of positive charge: the 'atomic nucleus'. The electrical attraction of opposite charges entraps the negatively charged electrons in a remote swarm around this compact central positively charged kernel.

Unlike the electron, which appears to be fundamental, the nucleus is a complex beast, a cluster of two types of particles that are nearly twins. One, the neutron, is electrically neutral, and about 2000 times as massive as an electron. The other, the proton, is almost identical to the neutron but for one essential attribute: the proton is positively charged. The number of protons in a nucleus determines its total electric charge and its position in the periodic table of elements—for example, helium, element number 2, contains two protons; nitrogen, the seventh, contains seven; and so on.

The positively charged nucleus is the source of electrical fields which seed the structure of matter. Most remarkable is the fact that the amount of positive charge on a proton precisely matches the quantity of negative charge on an electron. Any difference in the magnitudes of these balanced charges is too small ever to have been measured; they are the same to better than one part in a billion trillions. It is most likely that they are identical.

The balance is so perfect that the 15,000 coulombs of negative charge in those innumerable atoms' electrons are utterly neutralized. The attraction of opposite charges and repulsion of like charges governs the construction of atomic molecules, of crystals, and the shapes of much around us, not least ourselves.

These competing forces, rooted in the binary positive and negative sources of electric charge, cancel so precisely that at large scales it is the force of gravity that governs the motions of the planets and galaxies of stars. Yet were this balance spoiled at even one part in a billion trillion, electrical forces would dominate gravity, and our Universe would not exist.

Thanks to this electrical symmetry, nature has adopted a nearly perfect disguise, where the existence of vast amounts of electric charge is hidden deep within us. Yet, other than in their electrical charges, electrons and protons are quite dissimilar. The carrier of the quantum of negative charge, the electron, has no tangible extent and appears to be one of the basic letters of nature's alphabet. The carrier of the positive quantum on the other hand, the proton, has a measurable size. Discovered just over a century ago, it soon became clear that it would take ten thousand protons laid side to side to match the size of a hydrogen atom. So, protons are very small, but nonetheless each is at least a thousand times larger than an electron. What's more, it would take nearly 2000 electrons to counterbalance a single bulky proton. Is it not strange then that two such dissimilar bricks should be so perfectly symmetric in mirroring one another's electric charge?

Move on half a century, to 1968, and the enigma deepens. The proton's bulk and spatial extent relative to those of the fundamental electron suggested that the proton is not at the foundation of nature's structures. A growing suspicion that the proton itself contains a labyrinth of inner structure was confirmed with the discovery that protons are complex objects built from smaller pieces known as quarks. Like the electron, which appears to be a truly elementary particle, the lightest of all the electrically charged particles known as leptons, so too do quarks appear to be lightweight fundamental particles which seed the proton, neutron, and a host of other ephemeral strange, charming, and sometimes beautiful particles, collectively known as hadrons.

Leptons and quarks appear in many ways to be rather similar. The collectives of quarks—the hadrons—on the other hand are utterly unlike any leptons. Hadrons feel the strong nuclear force, which is a reason why protons and neutrons are found in the atomic nucleus, even though the positive protons repel each other. Leptons are in the remote outer reaches of the atom, blind to the strong force. The most common and stable examples of these two families, the electron and proton, are thus fundamentally very different, so why should they work together so well to form the Universe we inhabit? How are protons and electrons so perfectly related electrically?

Today, 2500 years after the discovery of magnetism and electric charge, the reason for matter's neutrality remains an enigma. Had this balance remained unsullied within each atom, it would have been very difficult to have discovered that electric and magnetic glue lurks within everything, giving matter its shape and form. Nature has buried its secrets deep but has not entirely hidden them, however. A temperature of a few thousand degrees is sufficient to break the electrical attraction binding electrons around the bulky central atomic nucleus. Even room temperature is enough to release one or two electrons. The ease with which electrons can move from one atom to another is the source of chemistry, biology, and life.

These insights only matured late in the nineteenth century, but awareness of electric charge and of magnetism is as old as science. For although the atomic seeds of bulk matter are electrically balanced, clues to the atomic architecture and its deep electrical structure are everywhere. It is a matter of recognizing them, and then interpreting what they mean.

The force within

Magnetism is a manifestation of electricity, and vice versa. Electricity and magnetism were imprinted into our surroundings from the beginning. Five billion years ago when the new-born Earth was a hot plasma of swirling electrical currents, these flows created magnetic fields. As the magma cooled to form what is today the world's solid outer crust, magnetism was locked into minerals containing iron, such as magnetite.

Today, the Earth's liquid core is still a terpsichorean frenzy of electric currents, which generate a magnetic field. This extends into the atmosphere and far beyond, invisible to our normal senses. But in spreading from its source in the molten core to the heavens above, it first permeates the Earth's crust. This is where it leaves a tangible imprint, evidence that there exists a force more powerful than gravity at work within the Earth whose influence extends very far.

Way back in the earliest Precambrian, four billion years ago, as the surface cooled, atomic elements accumulated in the strata. The most stable of these, iron, is today one of the most abundant elements in the crust. Igneous rocks formed from volcanic lava. These rocks have the property that in the presence of a magnetic field, their atoms of iron act like soldiers on parade as they themselves become magnetic. This is exploited in popular demonstrations where the magnetic field of a bar magnet can be made visible (Figure 1). Small filings of iron are first scattered on the surface of a table and then a magnet is placed carefully among them. Its magnetic field induces magnetism in the iron filings, turning them into thousands of miniature magnets. Each of these duly orients itself in the magnetic field, revealing how the direction of the magnetic force varies from place to place.

Figure 1 Magnetic field. A bar magnet induces magnetism in pieces of iron, revealing the presence of its magnetic field spreading from one pole to the other.

The bar magnet is a simple model illustrating what happens for the magnetic Earth itself. Earth's north and south magnetic poles are analogous to those of the bar magnet, our planet's magnetic field extending far into space. There are no iron filings out in space, but there are large amounts of iron ores in the hills, cliffs, and mountains on Earth. In some places, by chance, these magnetic clusters are quite extensive, as on the Isle of Elba and Mount Ida in Asia Minor, where large outcrops retain the magnetic imprint in rocks known historically as lodestone, now named magnetite.

There are legends how thousands of years ago in ancient Greece, a shepherd wearing leather shoes held in place by iron nails stumbled—literally—across magnetite when the powerful magnetism gripped the nails in his footwear. Whether or not a shepherd named Magnes discovered the eponymous rock, and if so whether it was in Magnesia, north of Athens, or on Mount Ida in Asia Minor, or even another Mount Ida in Crete, it is very likely that such experiences, if less dramatic than in the story, would have happened on various occasions.

Certainly, the power of magnetism would have been apparent ever since the Iron Age. Lightning is a flash of electric current which generates intense magnetic fields and magnetizes ferrous rocks. Smelting to retrieve the pure iron metal from these sources would have revealed their magnetic attraction. So, the phenomenon has probably been known for some 3000 years. Like the discovery of fire, that of magnetism probably arose in several places independently, all inspired by the natural magnetization of iron in rocks.

For magnetic rocks are ubiquitous. By the sixteenth century travellers recorded the best examples, from East India and the Chinese coast: 'Very massive and weighty, [the stone] will draw or lift up the just weight of itself in iron or steel.'[2] As knowledge of the phenomenon spread from Greek myth to Latin, and on to English, the names morphed into 'Magnes rock' or 'magnet'.

Amber nectar

Whereas magnetism left an imprint, electric charge manifested its presence only indirectly. There's any number of places where you can read that the Ancient Greeks discovered that rubbing amber with fur or silk would attract pieces of dry feathers. Amber's golden sunny colour led the ancient Greeks to use the word elector—shining one—to describe it. This evolved into the Greek 'electron', which was coined as the word to describe the basic unit of charge by Johnstone Stoney in 1881. Following discovery of the atomic constituent that carries negative charge, the word 'electron' has by today become associated as the name for this fundamental electrical piece of an atom. This testifies to the importance of amber in the history of electric charge but raises the questions: what is amber and why were the ancients rubbing it with a cloth?

For millions of years before humans emerged, coniferous forests covered large areas of the globe. Fossilized resin from these trees looks like solid golden honey. Amber's lustre made it a universal choice for jewellery back in the Stone Age, some 13,000 years ago. Amber glows when polished, and for prehistoric peoples the furs from dead animals were a handy way to brighten the object. No one anticipated that this chance conjunction of

[2] Robert Norman, *The Newe Attractive*, 1581.

amber rubbed by fur would reveal a piece of natural magic—the ability of amber to attract small pieces of skin, feather, or dried leaves and lift them from the floor, even make them dance in the air.

Amber's attraction of dust would have frustrated polishers who, after rubbing vigorously and sitting back to admire the result, then noticed dust and dry bits of dirt spoiling their handiwork. They would polish even more vigorously until a suitably bright jewel was achieved, but this was probably achieved only after they had added a bit of spit to their polish. Today we know that some plastics and shoe leather have a similar property when vigorously brushed, which is why the phrase 'spit and polish' was synonymous with attempts to shine shoes in preparation for special events. 'Spit and polish' became the name of the game for making jewels.

This magical power was known in Syria where weavers used amber for the end of the spindle. As the wheel spun, the whirling motion of the amber through the air electrified it, and it attracted pieces of cloth. The weavers dubbed it the 'clutcher'. Nor is this property restricted to amber; glass and jet (compressed coal) also have the remarkable ability of clinging to things after rubbing, and pith balls are especially sensitive to storing electrical charge. In modern society, plastic combs are more common than pieces of amber, so pull a plastic comb through your hair and then watch it attract small pieces of paper. On a dry day a rapid combing may even induce sparks, which are visible in a darkened room. This is a timid miniature example of a lightning flash, though without the sound effects, so it would have required a giant leap to associate these two phenomena. For over a thousand years, no one did.

What is electric charge?

Many years ago, in my schooldays, I bought a nineteenth-century book titled *The Reason Why—in Science.*[3] Comprising 1258 questions, all answered in 274 pages, it began with 'What is light?', 'What is heat?', and 'What are the attributes of heat?' Then came question number four: 'What is electricity?' The answer: 'Electricity is a property of force which resides in all matter, and which constantly seeks to establish an equilibrium. What electricity really is has not yet been discovered.'

Today an answer could be that electricity is the flow of electric charge, which in turn raises the question: 'What ultimately *is* electric charge?'

An honest answer is that we don't really know. Saint Matthew's adage 'Ye shall know them by their fruits'[4] is perhaps the most we can currently say. Electric charge is a concept invented to rationalize a wide range of what we call electrical phenomena. Historically, the first systematic study of these strange effects was the discovery that a pith ball brought in contact with a glass rod that had been rubbed first with silk would move away from the rod with a remarkably strong force. The phenomenon was clearly real and so was invented the idea that the pith ball had been positively electrified. Colloquially it was said simply that the ball had received a 'positive electric charge'. The amount of charge could then be determined by measuring the strength of the observed (repulsive) force.

[3] *The Reason Why—in Science*, edited by J. Scott, M.A., Sisley's Ltd, Makers of Beautiful Books (undated).
[4] Matthew 7:16 (KJV).

Likewise came the discovery that if an ebony rod is rubbed with cat's fur, the pith ball will be *attracted* to it. The phenomenon of a pith ball becoming subject to a novel powerful force appeared to be the same as before, except that repulsion had now become attraction. This switch in behaviour led to the idea that the ball has received a *negative* electric charge. The unambiguous phenomenon is that a pith ball can be put into either of these two states. It is then a matter of *definition* to say that it has received an amount of either positive or negative electrical charge.

All thinking about electrical matters, in particular electrical charge, starts with these two simple phenomena. Applying positive or negative electrical charge to two pith balls led to the discovery that balls charged with the same sign of charge—whether it be positive or negative did not matter—would repel one another, whereas when one was charged positive and the other negative, they would attract. That is the phenomenon that led to the rule that like charges repel whereas unlike charges attract. Why should that be? Ultimately, we don't know. Those are the phenomena; the rule of electrical attraction and repulsion in terms of positive and negative charges is an empirical observation.

The more vigorous the rubbing of the rods, the larger is the electrical force between the pith balls. This discovery fitted naturally with the idea that the rubbing produced the electric charge, the greater vigour causing larger amounts of electrical charge to be involved, leading to the increased strength of electrical force. The rule was found that the amount of force between two balls was proportional to the product of the charge on each one and that it died off as the inverse square of the distance between them. This was first established by the French scientist Charles Augustin de Coulomb in 1785. It is in honour of Coulomb that the unit of electric charge is named.

These phenomena were known long before the ideas of atoms and physical electrons were established. They naturally raised the question as to whether there was a minimum amount of electrical charge that one could attach in principle. Conceptually this is what inspired the American physicist Robert Millikan in 1909.

Instead of a pith ball, Millikan used the smallest possible spherical body which would retain its mass so that when released it would feel a constant gravitational force. In Millikan's celebrated experiment, the role of the pith ball was played by oil droplets about 1 micron in diameter which were blown out of an atomizer. The frictional effects of blowing the spray put charge on the drops. Millikan kept them in a stable atmosphere with no convection currents. In the absence of an electric field the drops fell due to gravity, but being very small and subject to the viscous drag of the air the rate was relatively slow, enabling it to be watched and measured.

Two metal plates, one connected to the positive and the other to the negative terminal of a battery, gave an electric field which could be turned on or off by means of a switch. The field goes directly from one plate to the other and the force acts on charged particles in the direction of the field. By switching the positive terminal to negative, and the negative to positive, the overall direction of the field, and hence the direction of the force on a drop, will also switch. The electric force either speeds up or slows down the drops depending upon the direction of the electric field and the amount of charge on the drop. By turning the field off, drops would fall to the lower plate and be lost, but flick on the field and some drops that had not yet reached the bottom would be arrested in flight or even begin to drift upwards towards the oppositely charged plate. By repeating this several times Millikan was able to remove almost all the drops until he was able to concentrate on just one for which the gravitational and electric forces balanced nearly enough that the drop could be held in suspension.

The clever idea was now to alter the amount of charge on the drop by illuminating it with ultraviolet light, or with the radioactivity from radium. The drop would now move under the action of the electric force with a speed proportional to the amount of this charge. He found that as the quantity of charge changed, the speed of the drop would be always two, three, or some other integer amount times a minimum, but never a fraction. The speeds are in discrete amounts, or 'quanta', because electric charge is quantized, coming in discrete amounts with a fundamental minimum magnitude.

Millikan won the Nobel Prize in 1923 for measuring the fundamental magnitude of electric charge and in his speech to The Nobel Foundation in 1924 he summarized the profound significance of the experiment. He said that one has literally seen the 'electron'—by which he means the basic quantum of electric charge—by measuring '(in terms of a speed) the smallest of the electrical forces which a given electrical field ever exerts upon the pith ball with which he is working and with the aid of whose movements he defines [electric charge] itself'. Furthermore, 'something which [we have] chosen to call [electric charge] may be placed on or removed from [our] pith ball only in quantities which cause the force acting upon it ... to go up by definite integral multiples of the smallest observed force'.[5]

[5] Robert A. Millikan, The electron and the light-quant [sic] from the experimental point of view. Nobel Lecture, 23 May 1924. https://www.nobelprize.org/uploads/2018/06/millikan-lecture.pdf.

By measuring thousands of drops, made of different substances of various sizes, and having them pass through gases whose viscosity and pressure were known, it proved possible to measure the electric charge not in terms of the speed that it gave to a given oil drop but in absolute electrostatic units: coulombs, in the modern language. Millikan answered our question 'what is electric charge' thus:

Electrons of both the positive and negative variety, are merely observed centres of electrical force.

Here by 'electrons' he means the fundamental unit of electric charge. He regards negative and positive as complementary, and thereby equal in magnitude. For him the mysterious asymmetry comes in the way the two are manifested by the fundamental particles (and recall that in 1924 when he gave this address, the particles named 'electron' and 'proton' were the only electrically charged fundamental particles then known). Millikan told the audience at the Nobel ceremony:

The dimensions of electrons may in general be ignored, i.e., they may both, for practical purposes, be considered as point charges, though, as is well known, the positive has a mass 1,845 times that of the negative. Why this is so no one knows. It is another experimental fact.

The interplay between 'electron' meaning unit of electric charge and also the particle carrying the negative form of that charge may be confusing to modern ears. That he recognizes there is a profound asymmetry at work is however clear. Therein is the basic puzzle facing us.

Millikan was alert to a possible caveat to his discovery, as he pondered:

Shall we ever find that either positive or negative electrons are divisible?

It is unclear whether by 'divisible' Millikan refers to the possibility of the physical electron and proton being built of smaller constituents, or to the equal quanta of negative and positive electric charge being multiples of some yet smaller quantum. As to an answer he concluded:

If the electron is ever subdivided it will probably be because man, with new agencies as unlike X-rays and radioactivity as these are unlike chemical forces, opens up still another field where electrons may be split up without losing any of the unitary properties which they have now been found to possess in the relationships in which we have thus far studied them.

At the centenary of Millikan's conjecture, this book reveals how much it has been answered.

Appendix: an electric shock

The smallest amount measured by Millikan corresponded to the amount of electric charge carried by a single electron—from here on 'electron' refers to the particle of that name. It is some 1.6×10^{-19} coulombs. With that number to hand, here is a brief diversion to check the shocking claims in the opening few paragraphs. Alternatively, you can take it on trust and skip this small section.

This is how I calculated them. You can check the individual numbers by going to a reliable source on the Internet or a textbook. Their combination is then mere arithmetic.

Equal volumes of gases, at the same temperature and pressure, contain the same number of molecules. This was first articulated by the Italian, Amadeo Avogadro, early in the nineteenth century. He computed what is known as Avogadro's number, which is the number of molecules of a gas needed for its weight to be the same as the molecular weight, measured in grams, a quantity known as a 'mole'. For example, for a gas of hydrogen molecules, chemical formula H_2, this would be 2 grams; for oxygen, O_2, this would be 16 grams, and for water vapour, H_2O, it would be 18 grams. Avogadro's number is huge, 6×10^{23}, and at room temperature and normal atmospheric pressure these weights of the various gases would each occupy a volume of 22.4 litres. A breath of air has a volume of about 0.5 litres, which is about one-fiftieth of the 22.4 litres required to make Avogadro's number. So, a breath of half a litre contains about one-fiftieth of 6×10^{23} which corresponds to 1.2×10^{22} atoms.

Air is dominantly nitrogen and oxygen in a ratio of roughly 4:1. Each nitrogen atom contains seven electrons whereas oxygen has eight. So, a half-litre mix of oxygen and nitrogen contains roughly 9×10^{22} electrons. Although each electron only carries a trifling amount of electric charge, some 1.6×10^{-19} coulombs, it is the vast number of them that accumulates the total of 15,000 coulombs. An average bolt of lightning transfers about 15 coulombs. So, the amount of negative charge you inhaled a moment ago is enough to ignite 1000 lightning bolts. As is the intake of breath in disbelief that perhaps you are taking right now.

2

THE NUCLEAR ATOM

Although the perfect balance of positive and negative within atoms has enabled nature to mask electric charge, even room temperature is enough to reveal the presence of flighty electrons in the atom's outer reaches. The source of the intense electric fields that fill the atomic inner space is the massive atomic nucleus at its centre. At room temperature the nucleus is deeply frozen into the atomic architecture; only at extreme temperatures, as in the heart of the Sun at tens of millions of degrees, does the nucleus become a leading actor.

Were that the whole story, the existence of the nucleus might have remained unknown, along with its role as the engine of the stars. Thanks to quantum uncertainty, however, even at room temperature evidence of the hidden nucleus can leak out, literally so in the form of radioactivity. As small breakthroughs can enable cryptographers to crack what appear to be impenetrable codes and gradually expose the meaning of previously indecipherable messages, so did the discovery of radioactivity give both the hint of an atomic code and the key to interpreting it.

The first inkling of radioactivity was so petty that it was almost missed. The chance discovery in 1896 that minerals containing uranium emit some mysterious radiation that can fog photographic plates showed there is a source of energy deep within atoms. The subsequent discovery of radium, a highly radioactive element, showed that atoms can emit energy seemingly without any prior stimulation, and for thousands of years. Two forms of radioactivity identified in the first decade of the twentieth century were named alpha and beta radiation. The particles comprising beta radiation are electrons, whose origin we will meet later. Initially the most significant discovery, and the key to unravelling the whole atomic architecture, was that the particles of alpha radiation are positively charged and about four times the mass of a hydrogen atom.

A key step was being able to observe atomic radiation directly. Alpha radiation consists of particles which naturally are known as alpha particles. When they hit a screen coated with zinc sulphide a flash of light is admitted. This phenomenon is called 'scintillation' and is a convenient way of observing the presence of alpha particles.

An alpha particle source emits alphas in all directions. To make a sharp ray of particles, the source was housed within an absorber containing a narrow opening. Any alpha particles that escaped through this opening formed a collimated, pencil-thin beam and when this hit the screen, a bright, sharp spot of light could be seen. This showed that in the absence of intervening forces or material, alpha particles travel in straight lines in accord with the laws of motion. When a powerful magnet was brought into play, however, the spot on the screen moved. This showed that a magnetic field affects their motion. A magnetic field deflects negative and positive charges in opposite directions; the deflection of the light spot in this case showed that the alpha particles carry positive electrical charge. Relative to the electron, an alpha particle has twice as much electrical charge (of opposite sign) and about 7300 times the mass.

Magnetic fields deflect fast-moving alpha particles; what is the effect of putting thin sheets of material in their way? Where the beam had previously left a small, bright spot on the screen, the spot became fuzzy when a thin sheet of mica was interposed. That there was any scintillation at all showed that alpha particles could penetrate the mica, but the fuzziness implied that in passing through the sheet, they were deflected slightly, in all directions.

That this happened was not unexpected because within the mica electrical forces hold its atoms together and these same forces would affect the motion of the electrically charged alpha particles. The surprise, however, was that the phenomenon was so appreciable. Measurements of the alpha particles had shown that they were travelling at about one-twentieth of the speed of light, giving them enormous energy for their size. This speed corresponds to some 15,000 kilometres every second, which in half a minute could get them to the Moon. Simple arithmetic shows that moving at that speed they could penetrate a thin sheet of mica in less than a nanosecond. That such flighty passage could be affected at all showed that the electric and magnetic fields inside mica must be huge, certainly much greater than anything known to technology in the world at large.

The brilliant New Zealander Ernest Rutherford, working at Manchester University in 1911, had the insight that these intense fields could be what held electrons in atoms. Effectively, Rutherford had found direct evidence of the fields that give rise to the structures of atoms and of materials at large. To investigate this further he fired alpha particles at heavy metals, such as thin sheets of gold leaf, using the scintillation method to detect what happened. To his huge surprise he discovered that alpha particles were sometimes deflected very violently, occasionally even being turned back in their tracks. He interpreted this as evidence that the positive charge within atoms, which is deflecting the alpha particles, is concentrated on a compact, massive central lump—the atomic nucleus. He further postulated that this positive centre is the source of the electric fields that hold the negatively charged electrons remotely, like satellites encircling a central attractor.

The conclusion from the experiment, that the nucleus is surrounded by electrons held by electrical forces, has survived for over a century and can be confidently included as a deep truth of nature. The picture of the atom as an electrical analogue of a miniature Solar System with negatively charged planetary electrons whirling around that compact central nucleus was very popular and is held by many even today. It is a useful image, and I shall make use of it, but it comes with a profound warning: it is wrong, or, at least, incomplete. The laws of Newton and Einstein, applicable to the motion of macroscopic objects, are not fundamental but have emerged from deeper laws that the innards of atoms make manifest. This is how.

Electronic waves

It is tempting to picture an atom as a miniature Solar System, where the nucleus plays the role of the Sun, and the remote electrons are like the planets. Whereas the force of gravity controls the motion of the planets, atoms are held together by the electrical attraction of opposite charges. So far, so good, but the analogy can be dangerous if stretched too far.

Atoms could not survive for a moment if they obeyed Newton's laws of mechanics. Atoms are very tiny, and the electrical force is much more powerful than gravity. Had electrons in atoms whirled around the central nucleus like planets orbiting the Sun and obeyed Newton's laws they would have radiated electromagnetic energy and spiralled into the nucleus within a mere fraction of a second. An atom, once formed, would self-destruct in a flash of light almost immediately. Matter, including you and me, would not exist. That we do, and atoms also, implies that something beyond Newton's classical mechanics is at work. The answer is: quantum mechanics.

Quantum mechanics is required to understand the stability of atoms and matter. Among its many implications, quantum mechanics dictates that an electron cannot go wherever it pleases in an atom. Its options are limited, like someone on a ladder who can step only on individual rungs. Electrons in atoms follow a fundamental regularity, each rung on their figurative ladder corresponding to a state where the electron has a unique amount of energy. When an electron drops from a high-level rung to a lower level, the difference in energy is radiated as light. It is the discrete levels of the rungs for energies of electrons inside atoms that lead to the discrete energies of the photons of light that are emitted. When these photons correspond to light in the optical range, their different energies show up as different colours. The resulting spectrum of individually coloured lines is like a fundamental barcode that identifies the atomic source.

But how does quantum mechanics constrain an atomic electron's freedom? In quantum theory, particles have a wave-like character. The idea that an electromagnetic wave can act as if composed of particles—'photons'—had been debated ever since Isaac Newton proposed the idea in the seventeenth century, but now quantum theory was giving it a fundamental rationale. And not just photons: quantum theory implies that any particle—the electron, for example—has a wave-like character. The most direct example is that of the electron inside an atom.

The golden rule of quantum physics is that the higher the momentum or energy of a particle, the shorter the length of the associated wave. Imagine the waves for electrons in atoms as if they were wobbles on a length of rope. Tether one end of the rope to some restraint and jiggle the far end. The faster you do this, the more energy you impart, and the shorter the wavelength becomes. As such, this is a visual analogy for the wavy electron in the quantum world.

Now imagine that electron in a circuit around the nucleus (Figure 2). In our model, the rope must be completed into a circle, like a lasso. To do this, the two ends of the rope must be at the same place in the oscillation cycle otherwise they will not match one another.

When coiled in a circle, for a wave to fit perfectly into the circumference, the number of wavelengths in the circuit must be an integer. Imagine this circle like a clock face. If the wave peaks at 12 o'clock, with a dip at 6 o'clock, the next peak will occur perfectly at 12. The wave fits into the circle. However, a peak at 12 followed by a dip at 5 o'clock would have its next peak at 10 and be out of time with the beat of the wave—out of phase in the jargon of physics. The wave will not fit.

The result is that the only possible pathways for an electron to survive in an atom will be those for which an integer number of wavelengths fits into a single circuit. A single wavelength corresponds to the electron being on the lowest rung of the ladder; two

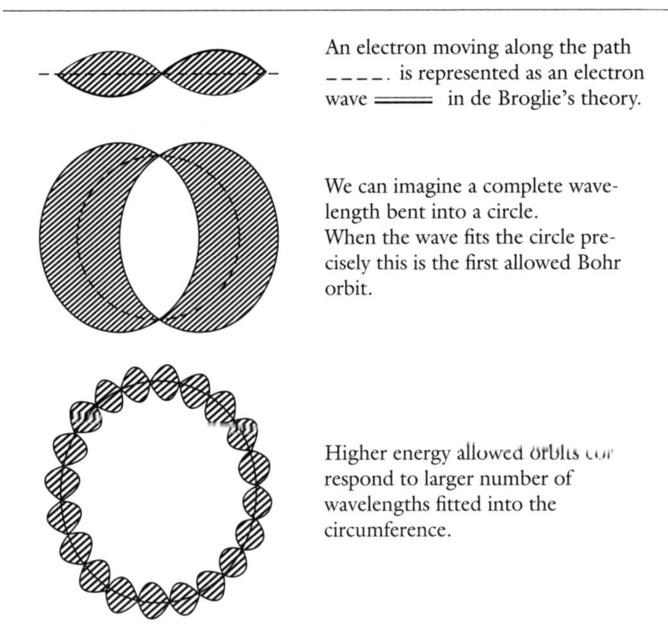

Figure 2 Electron waves and orbits in the atom. (a) An electron moving along the path - - - - - - is represented as a wave. (b) Imagine a complete wavelength bent into a circle. When the wave fits perfectly, this is the first allowed state and has the lowest energy. (c) Higher-energy orbits correspond to larger numbers of wavelengths fitted into the circumference.

wavelengths will find it on the second rung, three on the third, and so on. When an electron drops from a high-energy rung to a lower one, the difference in energies is radiated as light in perfect accord with the spectrum observed in simple atoms.

This is a beautiful example of the wave-like nature of electrons at least when inside atoms. The wave-like nature of electrons when they are free is seen when they pass through very narrow slits in an opaque screen. Beams of electrons passing through such slits turn out to be diffracted much like light waves. That electrons and indeed all fundamental particles take on wave-like characteristics is beyond dispute, but precisely what these waves refer to has been debated ever since the birth of quantum mechanics. I have nothing to add to this here, other than to say the concept of waves is fundamental to quantum theory and works empirically.

The nuclear code

When an atom is in an electric or magnetic field, its electrically charged constituents will respond to the electromagnetic force. Newton's laws of motion tell us that the amount of a body's acceleration, the change in its motion, is in proportion to the amount of force and inversely proportional to the object's mass. These rules are also true in quantum mechanics.

The nucleus of an atom carries more than 99.95% of its mass. In other words, less than 1 part in 2000 of an atom's bulk comes from the electrons whirling around the central nucleus. As a result, when electric or magnetic forces act on an atom, its electrons respond dramatically whereas the central nucleus remains relatively undisturbed. So long as an atom does not experience some extreme force that shatters the nucleus, it will be the nucleus that defines the atom's location. In turn, it is this static nucleus whose electric field entraps the electrons, and quantum mechanics that determines their energies.

When the atomic nucleus was first discovered, all that was known was that it is positively charged and very massive. Gradually it became clear that the nucleus has a complex structure of its own. The first insight was that the positive charge is carried by particles called protons and that each proton has the same amount of electric charge as an electron, but positive instead of negative.

In terms of mass, a proton is some 1836 times as massive as an electron, the two particles being so unlike each other that it is surprising that they happen to have such precisely matched amounts of electric charge. One might have expected that any arbitrary amount of charge could have been assigned to the proton. That this relatively bulky particle carries the same magnitude of charge though of opposite sign to that of the flighty electron, suggests that electric charge is quantized, in other words that it is some property that attaches to particles in discrete integer amounts. Why this should be so and what determines the amount are puzzles of modern physics.

The more protons that are gathered in the nucleus, the larger is its total electric charge. It is the amount of charge on the nucleus that determines the number of electrons that can surround it to form a neutral atom. These electrons determine the chemical properties of the resulting atomic element. Basically, it is the number of protons in the nucleus that determines both its charge and the resulting atom's place in the periodic table of the elements. For example, hydrogen, the lightest atomic element, has one electron surrounding a nucleus containing a single proton; the next, helium, has two, and so on up to the heaviest naturally occurring element, uranium, whose atomic nuclei each contain 92 protons surrounded by 92 electrons. In every case the electrical balance between nucleus and electrons is perfect, enabling atoms to be overall electrically neutral—at least, to the best accuracy that we can measure.

The Sun possesses 99.8% of the total mass of the Solar System, which is why it remains undisturbed while the planets orbit remotely around it. The case of the atom is even more extreme.

The mass of an atomic nucleus relative to the atom's electrons is some five to ten times greater than that of the Sun relative to the planets. The centre of mass of an atom is therefore found inside the nucleus. The nucleus is in effect like a static lump of positive charge spreading its electric tentacles uniformly into the surroundings, trapping the lightweight mercurial electrons.

The perfect counterbalance of negative electron and positive proton means that atoms even of heavy elements can exist with their electrical effects completely neutralized. It is relatively easy, however, to knock one or more electrons out of an atom, which creates a positively charged ion. Conversely, attaching electrons to atoms gives negative ions. The long-range effects where electric fields spread beyond a single atom, or where adjacent ions of opposite charge can mutually ensnare another, bind multiple atoms together to make molecules. Thus, although an individual atom may be overall electrically uncharged, two or more atoms encroaching closely upon one another can disturb their inner electrical activity and become ensnared in complex combinations.

The place of a given element in the periodic table is known as its atomic number, which, recall, is the same as the number of protons in its nucleus. Measurements of the relative masses of atomic elements, however, show that these numbers are not the same as the atomic number of the respective element. For example, helium with atomic number 2, whose nucleus contains two protons, is some four times the mass of a hydrogen atom with just one. All other elements show atomic masses that are larger than what one would expect if nuclei contained protons alone. The reason is that in addition to the positively charged protons, atomic nuclei can contain electrically neutral neutrons, a particle whose mass is almost the same as that of the proton. Adding neutrons to a nucleus increases the nuclear mass but leaves its charge, and hence the atomic chemistry, unchanged.

Two atoms whose nuclei contain the same number of protons will have the same chemical properties. The number of neutrons in their nuclei, however, may differ. One consequence is that atoms can have the same chemical properties but have different properties when it comes to radioactivity. Two such nuclei of an element with different numbers of neutrons are known as 'isotopes'—from the Greek meaning 'same place' (in the periodic table of elements).

Different isotopes of an element have different levels of stability. Protons mutually repel one another electrically, which adds to the overall electrostatic energy in the nucleus, increasing instability. Neutrons and protons mutually attract by the 'strong' force when they are in close contact. Neutrons have no electric charge, so adding neutrons to a nuclear clump adds more attractive centres without any electrical penalty and tends to increase stability. For this reason, the stable isotopes of elements tend to have more neutrons than protons in the nucleus. This phenomenon becomes more marked the further up the periodic table one proceeds. Whereas light elements such as helium, oxygen, and nitrogen can get by with an equal number of neutrons and protons in their nuclei, iron's 26 protons are stabilized by 30 neutrons, while at the extreme limit of stability in the periodic table, uranium's 92 protons require more than 140 neutrons.

If this were the whole story one would expect nuclei to prefer to be built of neutrons without protons at all. The reason why this is not so is because the neutron and proton, while both very massive compared to an electron, are not precisely identical in mass. An isolated neutron is slightly heavier than a proton. By Einstein's equivalence of energy and mass this means that a neutron at rest has a slightly greater mc^2—in other words, energy—than a proton. Every neutron that is added to the assembly increases this amount of energy in the collection; too many neutrons will eventually have added more energy than can be contained in a stable configuration.

Nature seeks stability and does so by reducing the energy in an unstable configuration. Energy is conserved overall, so it is more accurate to say Nature *redistributes* the energy, minimizing the amount in the previously unstable configuration and emitting the difference in some form. One example is beta radioactivity, which can arise when a neutron transforms into a proton. In this case, the energy difference appears in the form of an electron and a ghostly, almost massless neutral particle called the neutrino.

This process has two consequences. Inside the nucleus itself, a neutron has transformed into a proton. The nucleus contains the same number of constituents as before, but now has one fewer neutron and one more proton. The nucleus now has one more positive charge than before, but charge overall is conserved with the appearance of the negatively charged electron, which being blind to the strong nuclear force escapes from the nucleus. A change in the number of protons equates with a change in the chemical identity, which means that because of this beta radioactivity, the element has moved one place higher in the periodic table; in other words, nuclear transmutation has occurred. The second consequence is that the electron and neutrino are radiated out from the atom. The neutrino is so ghostly that it is hard to detect, but the electron with its electric charge cannot hide. The electron, or 'beta particle' in the historical jargon, is the detectable phenomenon that something has happened. It was the observation of the effects of beta radiation, such as fogging photographic plates, that first gave clues of the energy locked deep inside atoms and led to discovery of the atomic nucleus.

Beta radioactivity

The phenomenon of beta decay, where the charge of a nucleus changes by one unit, occurs when a neutron in that nucleus converts into a proton, or vice versa. When a neutral neutron becomes a positively charged proton, as in the example above, the electric charge overall is balanced by the emission of an electron. It is also possible for a proton to become a neutron, having given up its positive charge to a positron—the positively charged antimatter version of the electron. As our surroundings are made of matter, and as antimatter is destroyed when it touches its antimatter doppelgänger, the positron soon hits a stray electron and both disappear in a flash of light, annihilating in a burst of pure energy.

The fact that a neutrino is also produced was less obvious and took several years to establish. With meticulous measurements, the properties of beta decays producing positrons or electrons have been carefully studied. These experiments revealed that the change in energy of the nucleus in these beta decays does not equal the energy of the electron or positron. This charged particle is found to emerge with an amount of energy varying from almost nothing up to a maximum value. The interpretation is that the electron (or positron) is accompanied by an unseen particle, the neutrino. The total energy change is shared between these two particles, which explains why the energy of the observed electron (or positron) can vary.

None of the electron, positron, or neutrino pre-exist inside the nucleus. They are created by a force that has triggered the decay.

This force is known as the weak force, 'weak' reflecting the fact that its strength is much less than the strong attractive nuclear force that binds protons and neutrons in clusters, and also feebler than the electromagnetic force.

By the mid-twentieth century, the basic picture of matter was that it consists of neutron and proton inside the atomic nucleus, with electrons encircling this remotely, forming atoms. Neutrinos, which are electrically neutral siblings of the electron, make their appearance thanks to beta decay. Awareness of antimatter had also emerged, most notably with the discovery of the positron. A fundamental property of antimatter is that an antiparticle has the same mass as its particle analogue, and the same amount of electric charge but of opposite sign. Positrons therefore have positive charge of the same magnitude as the electron. Antimatter is destroyed when it meets matter. Positrons therefore play no role in building material atoms, nor do they have any relation with the bulky proton. That the amount of charge of a proton is identical to that of a positron is to the best of our understanding a fortunate phenomenon that begs explanation.

Key to beta decay is that the difference in charge of proton and neutron is equal to that between electron and neutrino. There is nothing that requires the electron and proton charges to counterbalance, or that both neutron and neutrino are neutral, however. This is key to our understanding of the behaviour of the fundamental particles and merits some careful explanation.

When a neutron converts to a proton, for example, the amount of its charge increases by an amount that we might call one positive unit. To preserve the overall charge, one negative unit of charge must somehow appear. Although matter has been transformed, from neutron to proton in this case, nothing has been overall created or destroyed. There was one nuclear particle to begin with, a neutron, and one at the end, a proton. This negative unit of charge is carried away by something that is overall not material, neither matter nor antimatter.

Today we know this carrier is a negatively charged W boson, in effect a massive electrically charged analogue of a photon. More on this later. The electric charge emerges in the macroscopic world carried by an electron, which is a material particle. As the W boson is neither matter nor antimatter, and as the sum total amount of matter and antimatter is conserved, what I just referred to as the neutrino is actually an *anti*neutrino. The material content of electron and *anti*neutrino cancel, the electric charge remains.

A piece of arithmetic now applies. Suppose the electron carries an amount of charge Q_e with the antineutrino carrying Q_a. The conservation of charge requires $Q_e + Q_a = -1$.

The charge of an antineutrino is opposite to that of a neutrino: $Q_a = -Q_v$.

Consequently, the conservation of charge equation becomes $Q_e - Q_v = -1$; in other words, the difference of the electric charges of electron and neutrino must equal the difference between the electric charges of neutron and proton.

There is nothing that requires the electron and proton charges to counterbalance as there is no combination that enforces $Q_e + Q_p = 0$. For example, beta decay could happen were the neutron and proton to have electric charges of, say, $-1/3$ and $+2/3$ relative to the proton's empirical amount, designated as $+1$. The neutrality of atoms enables beta decay to happen but is not required by it. That part of the charge puzzle remains to be explained.

3

THE ELECTROMAGNETIC FORCE

The Sun's gravitational field spreads throughout space in all directions uniformly, its intensity dying in proportion to the square of the distance. The electric charge of a static atomic nucleus gives rise to an electric field with similar properties. Within the confines of an atom, where the central nucleus is in effect a lump of electric charge at rest, this is in effect the whole story. If electric charge is in motion, however, the resulting electric current can give both electric and magnetic fields.

An electric charge that is orbiting or spinning produces a magnetic field. The direction of the rotation, clockwise or anticlockwise around some axis, will produce a magnetic field akin to that from a north or a south magnetic pole. Of course, what appears to be a clockwise rotation viewed from above, becomes anticlockwise from beneath. The model of a rotating charge gives a conception of a dipole magnet, acting like a north pole when viewed from one side but a south pole from the other.

On a grander scale the swirling electric currents inside the molten core of the Earth give rise to our planet's magnetic field. The north and south magnetic poles are where the field lines burst out from the planetary magnet into the surrounding space, where their effects extend for tens of thousands of miles.

Electricity and magnetism are intimately related. What I see as static charge, a passer-by will describe as being in motion relative to them. They will view that charge as an electric current. I will perceive that static electric charge as the source of an electric field, whereas they will see an electric current spawning both electric and magnetic fields. In general, electric effects mingle with magnetism, their perceived relative importance depending on your motion. This is a reason why relativity theory, true whatever your uniform motion may be, speaks only of their combination: the electromagnetic field. Physical phenomena, such as the repulsion between two positive charges or between two identical magnetic poles, are invariant but the mathematical description in terms of electric fields or of magnetic fields depends upon your motion.

An electric dynamo is a practical example of this profound intermingling. The dynamo contains two magnets, one that is stationary and the other rotating. The stationary magnet creates a powerful magnetic field; the rotating magnet distorts and cuts through that static magnetic field. The intertwining of these magnetic effects combined with motion gives rise to the flow of electricity.

In 1865 James Clerk Maxwell encoded all known electric and magnetic phenomena in a series of equations. These describe how electric and magnetic fields are spawned by electric charges and currents. By the end of the nineteenth century the most fundamental carrier of electric charge—the electron—had been identified. Maxwell's equations thus describe how the electric charge of a static electron is surrounded by an electric field and imply that shaking an electric charge will disrupt its electric field, leading to a burst of electromagnetic radiation. Something similar happens if a magnet is shaken and its magnetic field disturbed.

Measurements of the strengths of electric and magnetic forces help determine two quantities that appear in Maxwell's equations. Known by the tongue-twisters 'magnetic permeability' and 'electric permittivity', they are in effect measures of how easily empty space responds to a change in magnetic or electric fields, respectively. According to Maxwell's theory, electromagnetic radiation arises when an electric field transforms into a magnetic field, or vice versa. This seesaw is continuous, the result being a wave of electromagnetic radiation. The speed with which these waves travel is proportional to magnitudes of both electric permittivity and magnetic permeability and turns out to be the same as the measured speed of light.

The implication of Maxwell's equations, that an electromagnetic field oscillates and propagates as a wave travelling at the speed of light, produced an obvious further implication: light is an electromagnetic wave. Radio waves, X-rays, and the rainbow of visible light are all examples of electromagnetic waves distinguished merely by the frequency with which the fields are oscillating. What the human eye perceives as colours are its response to the different frequencies in a small range of the electromagnetic spectrum. This rainbow covers an octave, in the sense that the frequency of the violet extent of our vision is twice that of the red. Ultraviolet light consists of frequencies immediately higher than that of violet light, whereas frequencies lower than red light are known as infrared. Our eyes do not see infrared rays, but our skin can feel their impact as heat. At even lower frequencies are radio waves, whereas at the other extreme of ultra-high frequencies are X-rays and gamma rays.

Quantum theory implies there is a duality between waves and particles. In quantum field theory, the smooth legato up and down of intensity characteristic of an electromagnetic wave appears to be a staccato burst of individual intensities acting like a burst of massless particles: photons. Photons can appear and disappear. Strike a match or flick a light switch and vast numbers of photons will flash into existence. But photons are not forever. For example, those in sunlight are being absorbed by plants and their energy turned into life through the process known as photosynthesis. As you read this page, photons of light are entering your retina and vanishing, their energy triggering electrical signals in the optic nerve.

There is no place in Maxwell's classic theory where photons appear from nowhere and then disappear into oblivion. To understand photons from their creation to destruction requires quantum theory to be applied not just to particles, but to the electromagnetic field. Combining quantum mechanics, special relativity, and electromagnetic theory leads to quantum electrodynamics, the brainchild of Cambridge theoretical physicist, Paul Dirac. It is traditionally referred to by its acronym, QED, which is apt as QED is so successful that for nearly a century it has been the paradigm for quantum theories of force fields.

Portrait of an electron

The first step is to have a quantum description of a lone charged electron that is consistent with relativity theory. This story begins in 1928, when Paul Dirac combined the two great theories of the twentieth century, quantum theory and Einstein's special theory of relativity, to build a mathematical description of the electron.

Quantum theory describes very small things on the atomic scale, and Einstein's special theory of relativity deals with things moving very fast. Electrons are very light, and in electric fields can be accelerated to move very fast, so you need relativity to describe them. Dirac discovered that he couldn't achieve a mathematically consistent description by writing a single equation, however. Instead, his single equation had to bifurcate twice over before the mathematical logic was complete. Now Dirac had to interpret the physical meaning of the four linked equations.

The first doubling implies that the electron is more than a lump of charge. To behave consistently with the constraints of relativity, the electron must also act like a magnet, as if it has a north and a south pole. In other words, to behave in accord with Einstein's special relativity theory, an electrically charged electron must act like a magnetic dipole.

There was already empirical evidence for this before Dirac's mathematics stumbled on it. Electrons cannot move at random inside atoms but must follow specific paths constrained by quantum theory. This implies that they can exist only with certain discrete values of energy. Recall our analogy of a ladder, whose rungs represent the energy levels of the electron in the atom. When an electron drops from a high-energy rung to a lower one, the difference in its energy is radiated as light. It is because the values of these energies are discrete that the spectrum is not spread continuously like a rainbow but instead consists of a series of lines.

When atoms are placed inside a magnetic field, the character of the spectral lines can change. In some cases, a single line is found to split into a closely spaced pair. This was discovered by the Dutch spectroscopist, Pieter Zeeman, in 1896, before the quantum theory of electrons in atoms had been born. Today we understand the 'Zeeman effect' to show that what had previously appeared as a single energy level state for the electron is in fact two, their separation growing in proportion to the strength of the applied magnetic field. This implies the electron acts like a dipole magnet. Figuratively, its energy is slightly elevated or reduced dependent on whether its north or south pole is aligned with the direction of the magnetic force. This is analogous to the repulsion when two north poles are brought together, which requires extra energy to maintain, in contrast to the attraction, added stability, and lower energy when north and south are aligned.

While the first bifurcation had satisfactorily explained the electron's magnetic duality, the second bifurcation at first sight appeared to be nonsense. Whereas the first pair of equations satisfyingly described a magnetic electron with positive energy, the second bifurcation appeared to describe an electron but with negative energy. As Dirac's equation had been constructed to describe an electron completely free of all interactions with its surroundings, the idea of it having negative energy made no sense.

This apparent paradox concerned Dirac for some time before he realized that there was a way of reinterpreting its meaning. What at first appeared to be a solution implying negative energy for a negatively charged electron, Dirac realized could be reinterpreted as describing a positive-energy electron but with positive charge. No such thing was known, until in 1932 the positron was discovered in cosmic rays. Today we recognize the positron's discovery to have been the first identification of a particle of antimatter.

That Dirac's equation required the existence of antimatter in the form of the positron before humans had any empirical knowledge that such bizarre stuff exists, is an uncanny and remarkable example of the mysterious property of mathematics.

Mathematics appears to be able to describe a Universe in advance of humans being aware of the empirical reality for themselves.

Today several examples of anti-particles are known and many of these have the properties to which Dirac's equation applies. For our purposes the most relevant will be the antineutrino, the anti-matter partner to the neutrino. As the neutrino is the electrically neutral partner to an electron, so is the antineutrino paired with the positron. If the neutrino were to have an electric charge, then the antineutrino would have an equal amount of that charge but with the opposite sign. Empirically the neutrino has no charge, which implies that the antineutrino also is electrically neutral. You might well ask, what then distinguishes a neutrino from an antineutrino? One answer is that when neutrinos interact with atomic nuclei they convert into an electron, the electrical charge overall being balanced by a neutron in the nucleus converting to a proton. The analogous process for an antineutrino is that it converts into a positron, meanwhile in the nucleus a proton converts into a neutron, again preserving the overall amount of electric charge. In both cases the amounts of matter and antimatter are independently preserved.

Quantum electrodynamics: QED

Quantum electrodynamics—or QED—built on and extended Dirac's quantum description of the electron. In Dirac's equation the electron appears as a particle with charge e and mass m, at a point in a spatial void. In QED, this electron becomes a more complex entity. The reason is because it can interact with photons—the quantum particles of electromagnetic radiation. QED is an example of a quantum field theory, where in this case the field is the electromagnetic field. One implication of quantum field theory is that forces act by the exchange of particles. In QED an electron at one point may emit a photon that carries away energy and momentum. When this photon hits another charged particle, it will set that particle in motion by transferring its momentum to it. Newton's laws of motion imply that a change in motion is the result of a force, and when this is the result of electrical attraction or repulsion traditionally it is called the electromagnetic force. So, in QED the electromagnetic force between two electrically charged particles arises by the transient exchange of one or more photons. The photon is said to be the carrier of the electromagnetic force.

When we say that the electron is but a point of electric charge, what do we mean by a point? Experimentally the electron has been shown to be fundamental, in the sense of having no more basic constituents and no measurable size, at distances down to 10^{-18} metres. This is hard to imagine. Were we to enlarge an atom to the size of a football stadium, its nucleus would be no bigger than a pea. And 10^{-18} metres is a thousand times smaller yet.

It is possible that the electron has a size smaller than this, and if we could perform experiments with even better resolution we might find that the electron is composed of more fundamental particles, giving it an intrinsic size. To the best resolution available we have found no such size; we refer to it as point-like.

We can measure its shape. If the electron is indeed fundamental, it should be spherical because there is no preference for one dimension any more than another. Measurements show that it is remarkably spherical. If an electron were enlarged to the size of the Solar System, it would be spherical to an accuracy better than the width of a human hair.

The seemingly empty space around the electron is meanwhile teeming with an unseen swarm of ghostly spectators. These include the electric and magnetic fields surrounding the electron which in turn can fluctuate into pairs of particles and antiparticles—known as 'virtual particles'. An electron is inseparable from this cloud. A physical electron is more than the ideal entity that appears in Dirac's equation. Instead, what an experiment interprets as the electron's mass, for example, is the result of Dirac's naked electron interacting with its own electromagnetic field and with the cloud of virtual particles that surrounds it. The result depends on the resolution of the measurement.

An experiment to determine something even as straightforward as the mass or electric charge of a free electron is analogous to viewing something on a computer screen, where the image is made of pixels. At low resolution you see some level of detail, but if you increase the density of pixels, the picture becomes sharper. For example, if an image with a few pixels shows a river, representations built from larger numbers of pixels will reveal finer details of its swirling waters. Likewise, at low resolution, an electron's existence is revealed by a powerful electric field emanating from a source of electric charge. The electron itself can be discerned as no more than a fuzzy lump of electricity, which in our analogy fills a single pixel and is seemingly just beyond the limit of resolution.

To see what that haze of charge is really like, we must take a closer look. Suppose we increased the pixel density by 100, so that for every individual large pixel that was present in the previous case, we now have 100 fine-grain ones. Where before the electric charge was located within a single large pixel, we now find that it is situated inside the single mini-pixel at the centre, the 99 surrounding mini-pixels containing negatives and positives: the whirlpools of virtual electrons and positrons in the vacuum.

Now increase the density by another factor of 100. The charge of the electron, which previously filled a single mini-pixel, is now found to be concentrated into a central micro-pixel, which is surrounded by further vacuum whirlpools. And each of the whirlpools that had previously shown up in the image made of mini-pixels is revealed to contain yet finer eddies.

And so it continues. The image of an electron is like a fractal, repeating over and over at finer detail, forever. The electric charge of the electron is focused into a single pixel of an ever-decreasing size, surrounded by a vacuum of unimaginable electrical detail. QED takes account of these effects. If an experiment measures the properties of some electromagnetic process at one level of resolution, QED predicts how that same process will appear when studied at some different level of resolution.

What we refer to as the charge of the electron is in practice what we have measured at poor resolution. This might typically be a measurement of electric charge as done by Robert Millikan and Harvey Fletcher in 1909 and performed by students in university physics laboratories ever since.

That experiment involves determining the force of an electric field on ions, in effect on atoms stripped of one or more electrons. You don't know how many electrons are involved, but if the experiment is performed carefully the magnitudes are found to be integer multiples of a smallest possible amount. These different amounts correspond to the charges of an integer number of electrons, the smallest amount—or the difference between the discrete values—being that of a single electron. The result, 1.6×10^{-19} coulombs, refers to the measured charge of an electron when responding to the electric field in a macroscopic environment, in other words, at scales far larger than atomic dimensions.

What is measured here is in effect the result of the electric force acting on an electron complete with its cloud of virtual spectators.

To visualize what happens as resolution is improved it's perhaps easier to imagine a direct measurement of the coulomb force between two charges as the distance between them decreases to the atomic scale. As the two charges are moved nearer to one another, the amount of force between them initially grows in proportion to the inverse square of the separation, but when they encroach at atomic dimensions the effect of the virtual cloud of particles and antiparticles begins to have an influence. As the clouds surrounding the two test charges intermingle, the spatial dependence of the force will change subtly from an inverse square law. The interpretation of this in QED is that what we measure as the charge itself depends upon the resolution or energy of the experiment, and this causes the apparent deviation from Coulomb's inverse square law.

This change in strength has been confirmed empirically. When electrons and positrons annihilate at an energy of about 100 billion (giga) electronvolts—100 GeV—the product of their respective electric charges is seen to have increased by about 7%.[1] This increase has been confirmed to continue to the highest energies presently available at the Large Hadron Collider, of the order of 1 trillion electronvolts, 1 TeV. If we extrapolate the predictions to energies of the order of 10^{15} GeV, the product of the charges is predicted to have increased by a factor of about 3.

The message is that although charge is quantized, the magnitude of the quantum depends on the distance or energy scale at which it is measured. This implies that the nature of electromagnetic forces at extremely high energies, far beyond those yet accessible by experiment, will differ considerably from

[1] 1 eV is the amount of energy an electron gains when accelerated by a potential difference of 1 volt. It is equivalent to 1.6×10^{-19} joules.

those studied to date. This profound property will later have considerable implications for understanding the strong and weak nuclear forces and, eventually, the enigma of atomic neutrality.

Magnetic monopoles and the unit of charge

With his relativistic quantum equation for the electron, Dirac had successfully predicted the existence of a positively charged sibling and explained the source of an electron's (and positron's) magnetism. There remained, however, the puzzle of why electric charge is quantized. Dirac found a possible answer. He did so by focusing not on electric charge but on magnetism. For magnetism contains a profound puzzle.

We're all familiar with the idea of an isolated electric charge and it's easy to have a particle with a negative charge or positive one. Isolated magnetic charge however is harder to envisage. An electric charge can exist in isolation but a lone magnetic charge—known as a magnetic monopole—has never been found. The simplest known sources of magnetic fields are dipoles, the pairs labelled north and south. If you cut a magnet in half, you get two smaller magnets, each with north and south poles rather than single separate poles. Bring two like poles together and you feel intense resistance. Opposite poles, on the other hand, attract one another powerfully.

Maxwell's equations describe the behaviour of electric and magnetic fields; they have never been found wanting. One of these equations, known as Gauss's law for electric charge, says mathematically that an electric charge gives rise to an electric field whose 'divergence'—in effect, the intensity of its uniform spread into the surrounding three dimensions of space—is proportional to the magnitude of that charge. Maxwell's theory includes a similar equation, known as Gauss's law for magnetism. This says mathematically that a magnetic field has no divergence. 'No divergence' means, in effect, that there is no isolated source of magnetic charge. Instead, as another of Maxwell's equations implies, all magnetic lines of force emerging from a magnetic pole must turn around, or curl, and settle on a magnetic pole of opposite charge. If you spot one pole in a magnetic field, the opposite one will be nearby.

A bar magnet with its north and south poles is a famous example which makes the concept of magnetic field visible. To do this you need a magnet and lots of iron filings, as was seen in Figure 1. Their distribution gives a visual impression of the magnetic field, radiating out from one pole and curling round to return to the other.

The absence of magnetic monopoles was believed to be an axiom of Maxwell's electromagnetic theory. This changed in 1931, however, when Paul Dirac realized that quantum theory allows the existence of single magnetic poles. He came to this conclusion while thinking about a question that is near to our main theme. Why, Dirac wondered, does electric charge only come in discrete amounts?

The amounts of freely existing electric charge are empirically always a whole number multiplied by the size of a proton's charge. If we denote that amount by e, the amounts of electric charge on a piece of matter can be twice this ($2e$), three times ($3e$), or '(ne)' integer multiple, including negative ones such as $-e$ (as for a single electron), but never fractions such as $e/2$, or $5e/4$. Dirac's musings led him in 1931 to conclude that if a single magnetic pole exists anywhere in the Universe, this can explain why electric charge only exists in discrete amounts.

Dirac considered the following simple situation: a static electric charge and a static magnetic monopole, alone in the void. Maxwell's theory of electromagnetism allows such a situation in principle, the only uncertainty is whether nature makes use of magnetic monopoles. Maxwell's theory implies if this configuration can happen in practice, there will be a directional flow of electromagnetic energy and momentum around the pair. The amount of angular momentum in the flow is proportional to the magnitudes of the magnetic charge of the monopole and the electric charge of the electron that combined to create it.

There is nothing unusual in this. The directional flow of energy and momentum in an electromagnetic field is known as the Poynting vector, named for John Poynting who first derived it theoretically in 1884. Poynting gave a mathematical formula for the direction and the rate of transfer of energy—the power—when electric and magnetic fields combine. Dirac now considered the implications of quantum theory for the case where the electric field of a static electric charge mingles with the magnetic field of a monopole.

In quantum field theory, angular momentum is quantized. The magnitude of the angular momentum of the electromagnetic field in this case is proportional to the product of electric and monopole charges. Consequently, this product itself must be quantized. So, if even a single magnetic monopole exists in the Universe, the magnitude of the electric charge carried by an electron must be quantized.

Theoretical physicists love making bets about physics—who will win the Nobel Prize? Does the Higgs boson exist? (Those who bet against it lost.) Are there magnetic monopoles? For at least one theorist, the existence of magnetic monopoles is thought to be 'one of the safest bets one can make about physics not yet seen'.[2] Belief in the reality of magnetic monopoles might be a desperate way to accommodate our ignorance about a reason for atomic neutrality, for which charge quantization is probably necessary, though of itself hardly sufficient. Magnetic monopoles might be far too massive to be produced at high-energy accelerators, such as the Large Hadron Collider at CERN, or too rare to have shown up in cosmic rays, but according to Dirac's theory, discovery of just one magnetic monopole would be sufficient to explain why all electric charge is quantized.

It is tempting to conclude that the charge of an electron and of a proton each reflect the presence of this quantum. If this were so, then the negative quantum, carried by an electron, counterbalancing the positive quantum of a proton naturally leads to the electrical neutrality of the hydrogen atom. There is some metaphorical fine print in nature's lexicon, however: the charge on the electron (or the proton) is not the same as the quantum of charge. These may be the smallest amounts of charge found in isolation in the macroscopic world, but measurements at distances smaller than the size of a proton reveal that smaller amounts of electric charge exist. These are quantized, but in magnitudes that are one-third of the amounts carried by the proton or the electron. To understand the neutrality of an atom first requires us to understand the innards of the proton, where the basic quanta of electric charge appear to reside.

[2] J. Polchinski, *Int J Mod Phys*, A19, 145, 2004.

4

NEW AGENCIES

So far we have considered only electric charge. There are forces at work in nature which are spawned by various other forms of charge. In each case these charges have been given names, analogous to electric charge, and it is 'by their fruits' that we know them. However, there are some tantalizing similarities among the phenomena associated with these charges that point towards an as yet unrealized grand unifying theory that will explain the reasons for their existence and their properties. Today we recognize them as including what Millikan called 'new agencies'. Specifically, we now know that the 'electron'—in the sense of 'unit of electric charge'—can be subdivided.

Newton's law of gravity decrees that the magnitude of the gravitational force between massive objects is inversely proportional to the square of the distance between them and proportional to their masses. In the physics jargon, 'mass' is the 'charge' that is the source of the gravitational force.

Less familiar are the strong and weak forces that act in and around an atomic nucleus. The strong force, as we have seen, is felt by the nuclear constituents—protons and neutrons—and binds them together powerfully enough to resist the electrical repulsion among the multiple positively charged protons. The weak nuclear force disrupts atomic nuclei, causing them to decay by the process of beta radioactivity. Historically the route towards the theory of the weak force, and identification of the charge responsible, came before the strong force, so let's start with the weak force.

The weak left-hander

The simplest atom is that of hydrogen which consists of a single electron and a single proton, their negative and positive electric charges being precisely counterbalanced. After discovery of the neutron in 1932, for the next several years everything appeared quite straightforward. There were just three fundamental particles from which all matter could be constructed: electron, proton, and neutron. A fourth particle, the neutrino—a neutral nigh-massless sibling of the electron—was also inferred to accompany the electron emitted in beta decay.

If these are the fountainhead of everything, the simplest explanation for the electrical charges of electron and proton being opposite in sign and exactly equal in magnitude would be that electrical charge is some sort of external agent—a property of space perhaps—that is attached to matter in discrete amounts. Thus, if we start with electrically neutral matter (neutron and neutrino) then the addition of the unit of charge to the neutron or its removal from the neutrino will yield charged particles with exactly balanced amounts of charge. These are the proton and electron.

This was the sort of idea that Italian physicist Enrico Fermi had in mind in 1933 when he made the first theory of beta radioactivity. He assumed that a neutron decays and instantly converts at a point into a proton, an electron, and a neutrino. (Recall, this is now known to be an antineutrino but this detail was not yet understood when Fermi made his theory.) This was a very radical idea which had been inspired by discovery of the neutron the previous year, whereas the neutrino was still a hypothetical particle whose existence would not be confirmed until 1956. The equality of charges in Fermi's picture was rationalized by relating the electron and proton to neutral partners—the neutrino and neutron, respectively. Here we see the first emergence of the idea of families consisting of two members. One is the family of two leptons: neutrino and electron. The other is a family of two hadrons, strongly interacting particles: neutron and proton.[1]

Beta decay involves the transfer of energy and momentum from the nuclear particles to the emitted electron and neutrino. Newton's laws of mechanics tell us that the transfer of energy and momentum is the result of a force having acted. This concept remains at the foundation of quantum mechanics too. The force in this case cannot be the electromagnetic force, however, as that preserves the identity of the source of electric charge. For example, the electromagnetic force can cause a proton in a

[1] The idea of families persists today, but as we shall see later, with the neutron and proton family being replaced by down and up quarks. However, the idea of adding charge to fundamental neutral matter has been lost because the neutron is now known to be built from quarks which are themselves charged. The equality of electron and proton charges is therefore resurrected as a fundamental puzzle.

nucleus to emit a photon and lose energy while remaining a proton. In beta decay, however, a proton converts into a neutron, or vice versa. This forced transition between two siblings has no place in the theory of electromagnetism.

Thanks to Fermi's theory, the chance that a nucleus emits electromagnetic radiation relative to its chance of undergoing beta decay could be calculated and from this the strength of the novel force relative to the electromagnetic force could be determined. The chance of beta decay occurring turned out to be relatively small, a result that led to the force involved becoming known as the 'weak' force, the moniker reflecting its strength relative to that of the electromagnetic force.

Fermi's picture has been subtly modified over the subsequent 90 years, but the essential premise—that beta decay involves transition between two siblings—survives. Apart from its trifling strength, many features of the weak force appeared to be common to the electromagnetic. The ways that particles transfer energy and momentum from one to another, for example, are similar, the 'weak' and electromagnetic forces appearing like two manifestations of a single reaction. Fermi had exploited this when he constructed his theory of beta decay by analogy with electromagnetism. However, in 1956 came a discovery that shattered this naïve symmetry.

An electron can 'spin' in one of two configurations. It doesn't actually spin literally, but this is a convenient way of visualizing the magnetic duality revealed by the first bifurcation in Dirac's quantum equation. These are traditionally referred to as clockwise or anticlockwise, where like a conventional corkscrew the clockwise is referred to as right-handed and the anticlockwise is left-handed. In a mirror image, these get reversed. The concept of mirror image plays an important role in understanding the properties of particles and their interactions. The matching between a real and mirror image in particle physics is referred to as 'parity' symmetry.

Here is an example of mirror symmetry in particle physics. Imagine a left-handed electron feeling an electric or magnetic force and being deflected in some direction, to the left let us suppose. A mirror image of this would show the corkscrew rotating in the same direction but the motion of the electron will be reversed, so what was previously a left-handed electron, now appears in the mirror as a right-handed one. The deflection will also be reversed, being now by the same amount but to the right. Mirror symmetry, parity, says that in the real world this process where the right-handed electron bounces right has the same probability and characteristics as that when a left-handed electron bounces left.

When an electron is scattered by the electromagnetic force, this symmetry is indeed confirmed. Empirically, the probability for the left-handed electron to bounce by some angle to the left and lose some specific amount of energy in the process is the same as that for the right-handed electron losing that energy when bouncing through the corresponding angle to the right. In 1956, however, came the surprising discovery that processes controlled by the weak nuclear force do not exhibit parity symmetry. In other words, the mirror image shows a process that has different characteristics than those found in the real world. Indeed, it is even the case that the mirror world reveals processes that do not occur empirically at all. The weak force is said to violate parity, maximally.

One among many examples is when neutrinos are involved. For present purposes we can imagine the neutrino as like an electron whose electric charge and (most of its) mass have been removed. Fermi's theory of beta decay (Figure 3), where a neutron converts into a proton, electron, and neutrino, effectively at the same place, implies that a pre-existing neutrino can hit a target and pick up electric charge. For example, a neutrino hitting a neutron will transfer electric charge so that the neutron converts to a proton and the neutrino converts to an electron. Electric charge has been preserved overall but swapped around between the particles involved.

A neutrino however is more subtle than merely an electrically neutral electron. In practice a neutrino can only spin in one direction, conventionally referred to as left-handed. A mirror image of this neutrino would reveal a neutrino spinning right-handed. There are no known right-handed neutrinos in nature. If you were to see a right-handed neutrino converting to an electron, you could be sure that you were looking at the mirror image of a real process, not the actual thing itself.

The violation of mirror symmetry here is total. It is not simply a case of the chance of a right-handed neutrino interacting

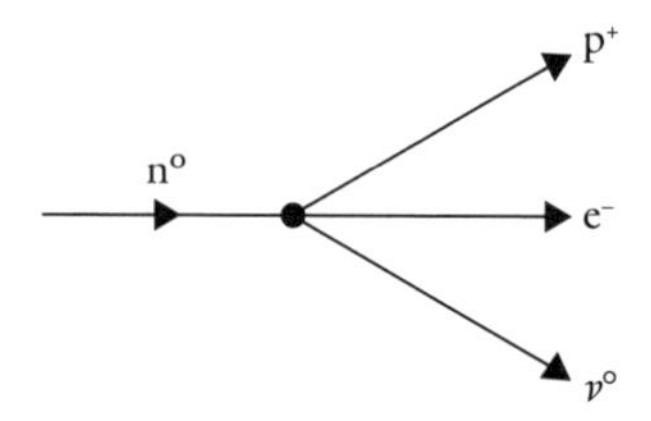

Figure 3 Neutrino picks up electric charge. Fermi's model of beta decay. (a) In Fermi's model, a neutron denoted n^0 turns into a proton p^+, electron e^-, and neutrino ν^0 at a single point in space. The superscripts denote the amount of electric charge that each particle has relative to a proton, and the sign denotes if it is positive or negative.

with matter being smaller than that of a left-handed one; empirically there is no such thing at all. This maximal violation of parity symmetry was one of the most remarkable discoveries about the weak interaction, back in 1956. Much as when constructing theories of particles and forces we insert empirical values of masses, for example, into the equations, so too do we include parity violation to ensure the accounting is correct. However, we have no fundamental explanation as to why nature has chosen this sinister path. Parity violation is currently a mystery no less profound than that of the atom's electrical balance.

As the weak force can add charge to a neutrino, converting it into an electron, so is it possible sometimes for an electron to have its charge annulled and convert into a neutrino. At least, this is possible if a high-energy left-handed electron feels the weak force because that force can turn a left-handed electron into a left-handed neutrino without penalty. A right-handed electron can bounce off the target and remain a right-handed electron, but it cannot transfer charge and convert into a right-handed neutrino. Only a left-handed neutrino takes part in the charge transfer weak interaction.

Traditionally we say that the weak interaction here is left-handed, at least when particles are involved. In the world of antiparticles, however, left and right are swapped. A right-handed neutrino may not exist in the real world, but a right-handed antineutrino does. Where a left-handed neutrino can convert into a left-handed electron, so does a right-handed antineutrino convert into a right-handed positron. Apart from the swap of matter into antimatter and left into right, everything else about these two processes is the same. In summary, all the evidence is that the weak interaction is intrinsically left-handed when acting upon the fundamental leptons—the electron or neutrino, for example—whereas for the corresponding anti-particles it is their right-handed forms that are involved.

The mirror cracked

The pion is a hadron which is produced in abundance when cosmic rays hit atomic nuclei in the upper atmosphere, or when beams of protons hit targets in a laboratory. It has no spin, is unstable, and occurs in three forms: one electrically neutral, the others charged positive or negative in the same amount as an electron or a positron. The electrically neutral form can decay into two photons, whereas the charged forms can decay to a neutrino and a muon. The muon is a heavy version of an electron, whereas neutrinos are electrically neutral analogues of either of these. We say the analogue of an electron is an 'electron-neutrino' and analogously the uncharged version of a muon is a 'muon-neutrino'. Unlike the electron and muon, the neutrinos were for a long time thought to have no masses but are now known to have them, albeit trifling compared to those of the electron and muon. For our purposes it will be sufficient to suppose that neutrinos are massless.

Each of the neutrino, muon, and electron is a 'fermion'—a particle which has a spin magnitude of 1/2 in units of the fundamental measure of quantum physics, known as Planck's quantum of action. If parity symmetry applies, each of these fermions can spin with the axis of spin—the corkscrew, in effect—either parallel or opposite relative to their direction of motion. When a pion decays, the conservation of linear momentum implies that the muon and neutrino will fly off back-to-back relative to the direction of motion of the pion. Their spins must also conspire to cancel one another, in memory of their parent pion having had none.

If we sit next to the pion so it is effectively at rest, then the muon and neutrino will move off with the same momentum but in opposite directions. The original spin being zero means that the spins of the muon and neutrino must counterbalance. So, if the neutrino's spin is left-handed—meaning projected opposite to the direction of its motion—then to annul that spin, the muon moving in the opposite direction must also be left-handed. (See Figure 4.) The same remarks would hold true were the neutrino right-handed in which case the muon would also be right-handed.

If parity is a good symmetry, then each of these situations is equally likely. Empirically however, only one of them occurs because parity is maximally spoiled. Neutrinos produced this way are always left-handed; antineutrinos being right-handed. This in turn constrains the spin orientation of the muon.

If you could measure the spin of the neutrino, you would immediately have evidence for the violation of parity symmetry, assuming of course that you knew also that it was indeed a neutrino and not an antineutrino that you were studying. What we

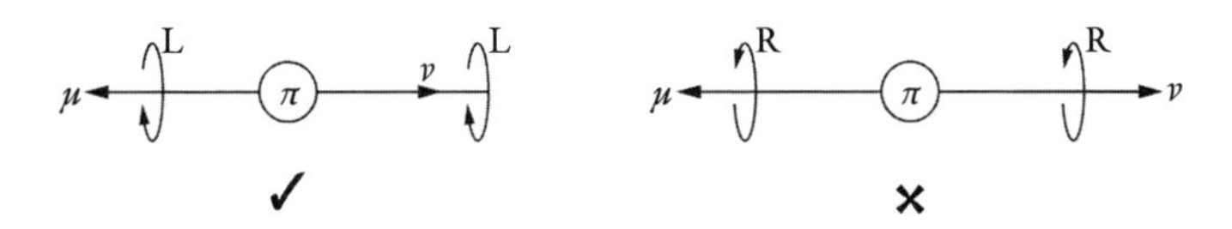

Figure 4 Pions, muons, and parity. When a spinless pion decays, the net spins of muon and neutrino must point in opposite directions. As they also travel in opposite directions, this means the direction of their corkscrew must be the same. Two possible situations can arise, shown in panels (a) and (b). If parity symmetry is respected, each of these two possibilities is the same. Empirically, only one occurs.

can measure is the electric charge of the muon. As the neutrino is neutral, the muon will carry the electric charge previously owned by the pion.

The electron with negative charge is conventionally identified as matter, while the positron is antimatter. Analogously, the negative and positively charged muons are identified respectively as matter and antimatter. A pion is neither matter nor antimatter, or if you prefer, it is an equal mix of each.[2] When a pion decays, the result will contain equal numbers of matter and antimatter particles. So, when the positive pion transfers its electric charge to the positively charged muon, which is antimatter, it will be accompanied by a neutrino—matter. Similarly, decay of a negative pion produces the negative muon—matter—accompanied by an antineutrino.

Neutrino and antineutrino do not themselves carry electric charge, but they do metaphorically carry some property constraining the neutrino to pick up negative electric charge, as when converting to an electron, whereas an antineutrino converts to the positive positron. For lack of any deeper insights, we refer to this as 'flavour'. Flavour appears to be another variety of charge, one of Millikan's 'new agencies'. As electric charge is the source of the electromagnetic force, so is flavour charge the source of the weak force. In constructing our description of particle forces and patterns we link the neutrino and negatively charged electron in a pair. Similarly, we place the antineutrino and the positively charged counterpart in another pair.

Although it is hard to measure the properties of a neutrino or antineutrino directly, we can infer some of these by observing the

[2] Hadrons are made of more fundamental constituents—'quarks'—see chapter 5. A pion is made from a quark and an antiquark. The positively charged pion, for example, is made of an up quark and a down antiquark, the negative version being a down quark and an up antiquark.

associated charged partner. For example, in the decay of the pion we can infer how the neutrino is spinning if we can measure the spin of the charged muon and apply the principle that the total angular momentum (in this case, spin) is conserved. It is by such measurements that we can demonstrate that mirror symmetry, parity, is maximally violated in such processes.

The muon is itself unstable, however, and decays in about a millionth of a second. If the muon was negatively (or positively) charged it decays to an electron (or positron) together with a neutrino and an antineutrino (Figure 5). If parity symmetry applies, there will be no fundamental preference for the electron (or positron) to emerge in one hemisphere rather than its mirror opposite.

In 1957, three American physicists in New York, Richard Garwin, Leon Lederman, and Marcel Weinrich, made the measurements and confirmed that parity is maximally violated in this process. Pions generate muons that are spinning like subatomic corkscrews, which in turn decay producing electrons. The experimenters then observed the direction of the electrons

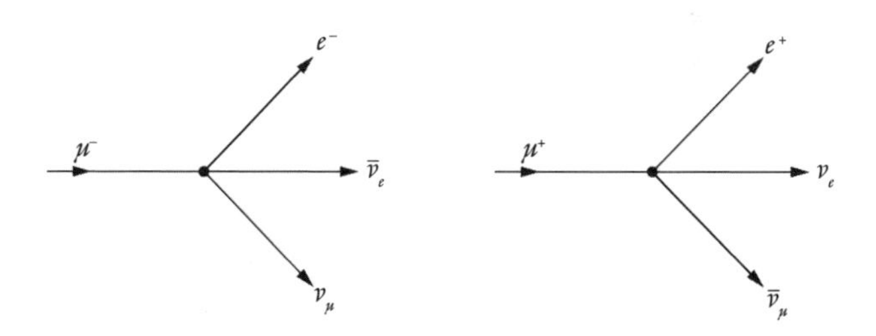

Figure 5 Muon decay. A negatively charged muon decays into an electron, a muon-neutrino, and an electron-antineutrino. A positively charged muon decays into the corresponding antiparticles: a positron, a muon-antineutrino, and an electron-neutrino.

emitted when the muons decayed. The basic idea initially was to stop the muons in a block of carbon and see how many electrons decayed in the forward hemisphere and then rotate the detector so that they could measure how many decayed in the rear hemisphere. If parity symmetry is satisfied, this would be like looking in a rear-view mirror and the chance of finding electrons from the decays would be the same. If the numbers were markedly different, however, mirror symmetry would fail.

That was the basic idea but is easier said than done. First: make your muons.

This is relatively easy, as by smashing a beam of protons into a target, large numbers of pions emerge. The momentum of the protons' impacts throws the pions forward. These pions decay in a fraction of a second, producing a beam of muons that are aligned along the deceased pions' direction of motion. Fortunately, the orientation of a muon's spin survives the effect of electromagnetic forces thanks to a beautiful fact that for fundamental fermions the rate of deflection of the particle's motion is in precise synchrony with the rate that the orientation of its spin rotates. The result of this balancing act is that if a muon begins its journey through the field with its spin pointing in the direction of its motion, say, it will remain in that configuration throughout. The same is true if the muon's spin is pointing in the opposite direction. The effect is to steer the muons without changing the orientation of their spin relative to their direction. In other words, if muons are born spinning right-handed, they remain right-handed. The same preservation of spin is true for left-handers too. Each muon's spin is effectively anchored along the direction of its motion.

A muon is unstable and survives, at rest, for a mere 2.2 microseconds. In motion, thanks to the effects of time dilation in special relativity, it survives longer, but decays eventually nonetheless. What remains is to detect the electron (or positron) produced in the muon's decay. If parity symmetry holds true, the particle is equally likely to emerge in the forward or backward directions. However, if the electrons or positrons were to emerge more in one direction than the other, parity symmetry would fail.

This seemingly straightforward measurement contains a hidden danger. Orienting a detector first to look one way and then the other could inadvertently break the mirror symmetry. For example, if the distance of the detector from the muons in the two cases differs, this would alter the yield of detected electrons because of a failure of mirror symmetry in the set-up, and not necessarily in the fundamental process itself (Figure 6a). The American team's inspirational idea was to leave the detector alone and instead to loop the muons through a circle by means of a magnetic field. Key to this trick is that a spinning electric charge

Figure 6 Measuring electrons spawned by muons. (a) The top situation detects forward decays, while the lower detects backward. This is only a test of parity violation in muon decay if the two detectors are the same distance from the decay point. We have exaggerated here to show this is not the case. (b) The detector stays fixed while the direction of the muons is changed. Statistically the position of muon decays can be averaged out and a cyclic variation in intensity which matches the rotation time of the muon beam confirms a fundamental parity violation.

acts like a magnet and when in a magnetic field, the direction of its spin will turn like a compass needle. Whereas a compass needle settles in one direction, pointing to the north–south polar axis, the mechanical forces acting on a magnetic muon make its spin rotate continuously. This it does as the muon takes a small loop, the spin being locked along its flight path for the reasons just explained.

That way it would be possible to compare the yield from decays of muons whose spins were oriented towards the detector to those halfway round the loop whose spins would be in the opposite direction (Figure 6b). As empirically parity is maximally violated in both the production and decays of the muons, this produced a measurable asymmetry in the intensity of electrons.

In summary, this is what has happened. Nature is very accommodating as parity is maximally violated both in the production of the muons and when they decay. Parity violation in their production causes the muons to appear with their spins completely polarized (oriented in but one of the two possible configurations). Maximal parity violation in the decay has the consequence that muons whose spins are oriented in the direction of their motion will produce electrons dominantly along that direction (let's suppose), whereas muons spinning in the opposite sense will produce electrons more in the backward direction. The effect in the experiment of making the muons' spins rotate will be to vary the intensity of electrons heading forwards, towards the detector. That is indeed what they found.

The whole experiment was completed over a single weekend. Since then, innumerable experiments have confirmed parity violation and determined its properties.

Who ordered that?

The discovery of the muon in 1936 opened a mystery that remains unsolved. The muon appears naïvely to be simply a heavier version of an electron, but nine decades of research have demonstrated that the muon carries some intrinsic 'muonness' not shared by the electron. The muon is some 207 times as massive as an electron. If it were no more than a heavier version of the electron, a muon would be simply an electron with extra energy—thanks to Einstein's equivalence of energy and mass expressed through $E = mc^2$. The electrically charged muon would then be able to shed that extra energy by emitting electromagnetic radiation—photons—stabilizing itself as the lighter electron. In decades of experiments, however, not a single example of such a transition has been observed. It is in that sense that the muon appears to carry some special quality that the electron does not share, and which is preserved. We can give it a name, 'muon-flavour', and contrast it with 'electron-flavour', but beyond that categorization we have no real insight as to what this distinguishing property is.

This concept is more than an ad hoc way to rationalize the absence of muon radiative decay because it is manifested in other processes also. As we have seen, a muon decays to an electron by emitting a neutrino and antineutrino. The neutrino being accompanied by an antineutrino preserves the overall number of particles. At the start there is one particle, a negatively charged muon; at the end there is one negatively charged particle, the electron, while for the neutrino and antineutrino matter and antimatter counterbalance.

The neutrino and antineutrino that appear are not simply particle and antiparticle of one another, however. At the start we had a particle charged electrically negative which was also charged with 'muon-flavour'; the electrically negative particle at the end is charged with electron-flavour. The muon and electron flavours—charges—are separately preserved because there are two varieties of neutrino too. If the electron-neutrino is defined to carry a positive amount of electron-flavour, then the antineutrino analogue will carry the same negative amount. Similar remarks will apply to the muon-neutrino and antineutrino. (This was illustrated in Figure 5.)

So, we see that both muon-flavour and electron-flavour charges are conserved in this process thanks to the production of a muon-neutrino and an electron-antineutrino. At the start, there is one amount of muon-flavour and none of the electron variety; at the end, the muon-neutrino carries the muon-flavour whereas the electron and the electron-antineutrino have the net electron-flavour mutually cancelling.

The result of all this is that the muon and its associated muon-neutrino form a left-handed pair in weak interactions precisely analogous to the way that an electron and its associated neutrino do. All experiments to date confirm that the muon and muon-neutrino take part in weak interactions in an identical way to the electron and electron-neutrino. Adding to our confidence in this observation, as well as adding to the intrigue, is the fact that a third variety of these particles has been discovered. The negatively charged tau lepton is the electrically charged clone of the electron and muon except that it has a massive 3500 times the heft of the former and carries its own unique 'tau-flavour'. Partnered by a neutrino, known as a tau-neutrino, we have here a third example of the flavour doublets of leptons.

Other than the fact that these various particles have different masses, all their behaviours in the weak interaction appear to be identical. The electron-flavour, muon-flavour, and tau-flavour charges generate weak forces the same, yet as they do so, somehow in the background these three distinct properties also lurk. What their distinguishing character is we do not know. Although flavour is a charge, in that it is the source of the weak force experienced by flavoured particles or antiparticles, the discrete natures of 'electron'-, 'muon'-, and 'tau'-ness are as yet obscure. What differentiates these three properties does not appear to be analogous to electric charge, however, because it does not seem to be the source of some further novel agency or force acting on electron, muon, and tau in disparate ways. However, remembering the caution of Millikan a century ago, we might admit that the possibility remains that there could be further forces so feeble as to have yet passed notice, and the existence of distinct electron-, muon-, and tau-flavours may be a clue to their existence.

5

QUARKS

By 1950, the naïve hope that all matter can be understood in terms of proton and neutron, the electron, and the still hypothetical neutrino was rapidly disappearing. Large numbers of particles were being discovered, initially in cosmic rays and later in experiments at earthbound particle accelerators. Some of these, such as so-called K and Lambda particles, had singular properties—typically being produced in pairs but decaying in isolation. They became known as strange particles. This led to the concept of a new form of charge, dubbed 'strangeness'.

Whereas all forces conserve electric charge, strangeness is conserved by the strong and electromagnetic forces but not the weak force (Figure 7). For example, when the strong force spawns a particle with positive strangeness it also produces another particle carrying negative strangeness, preserving the net zero strangeness overall. An example is the production of a Lambda with one

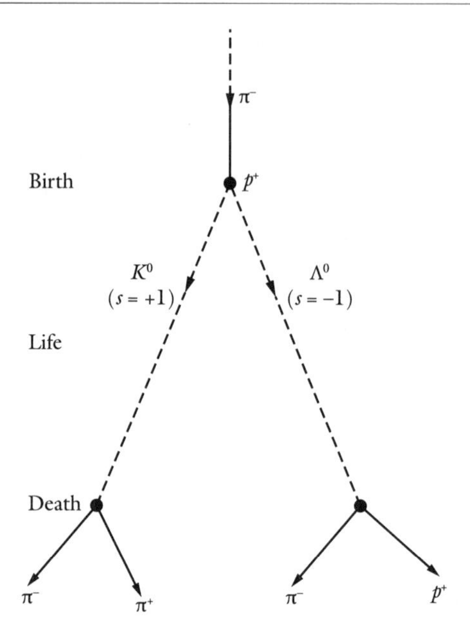

Figure 7 Strangeness in strong and weak interactions. (a) The strong force produces a particle K with strangeness +1 along with one (Lambda, Λ) with strangeness −1, conserving the total strangeness at 0. (b) The weak force does not conserve strangeness. A strange particle (such as an electrically neutral K, denoted K^0) decays to two pions (π^-, π^+). The pions have no strangeness. One is electrically positive and one electrically negative, conserving the electric charge, but strangeness is not conserved.

unit of negative strangeness in association with that of a K with one unit positive. The weak force, by contrast, can destroy strangeness. The action of the weak force thereby enables isolated strange particles to decay, whereby the individual K and Lambda convert into particles without any strangeness.

Strange and numerous though these particles were, there was a very clear simplicity with regard to their electric charges. Almost all particles known then, and since, occur with electric charges either positive or negative in the canonical amount, or of zero. In one case a particle with no strangeness but with two units of positive electric charge showed up. This is known as the Delta resonance. This created an intriguing asymmetry because there was no analogous particle with electric charge –2. That puzzle apart, every particle ever discovered has its electric charge quantized in the same way, one or two times that of a proton. Likewise, strangeness is quantized in positive or negative integer amounts: 0, 1, 2, or even 3. No fundamental particle with electric charge 3, or strangeness 4, has ever been found. Both electric and strange charges are quantized, but the amounts are limited.

From this veritable zoo of particles, the first task was to collate their properties and order them like a taxonomist. This quickly led to the identification of groups of particles forming families. For example, the proton and neutron were clearly two of a pair, distinguished basically by their electric charge. Strange particles were discovered with electric charges +, –, and 0, whose properties are like those of the proton and neutron apart from their strangeness and slightly heavier masses. These are known as Sigma particles. Two more particles were discovered, known as Xi, with analogous properties to these, except that they are yet heavier and carry two negative units of strangeness, their electric charges being negative or zero. One final member of this family exists—the Lambda baryon—with one negative unit of strangeness and no electric charge. This made a total of eight in all.

Further families of eight with this same set of correlations between electric and strange charge were found, along with families of ten. Occasionally, lone outliers—'singletons'—were found with properties that linked them like satellites to a family of eight. The discovery of these families containing one, eight, or ten members brought order to the large number of particles known as hadrons. The first clue to explaining why these patterns formed came from a branch of mathematics known as group theory, which deals with relations or 'symmetries' between similar objects. The symmetry group in this case is known as 'SU3'—pronounced 'ess-you-three'. The S and U have interest primarily to mathematicians; it is the '3' that is of fundamental interest to understanding the origins of the families of hadrons.

We don't need to know the mathematics of SU3. For our purposes it is sufficient to say that some families of eight together with a lone singleton arise because in the mathematical scheme 3×3 can be mathematically separated into $1 + 8$.[1] The families of ten turned out to be associated with two independent families of eight, these all accompanied by a lone singleton. These families arise in the mathematical scheme because $3 \times 3 \times 3$, giving a total of 27, is the same as $1 + 8 + 8 + 10$.

Hadrons come in two varieties: baryons, particles like the proton and neutron with half integer spins; and mesons, such as the pion, with integer spins. Mesons were found always to form families of eight and one and were understood to arise from the $3 \times 3^*$ feature of SU3, whereas baryons always fitted with the combinations of one, eight, and ten arising from the $3 \times 3 \times 3$ feature of the theory. As numerous particles continued to be discovered which fitted into these families, there was no doubt that there was something behind this taxonomy, but what?

[1] Technically $3 \times 3^*$ where 3^* refers to an anti-triplet in SU3.

The quirks of quarks

Murray Gell-Mann, the American theorist who first developed this mathematical scheme and whimsically named it the 'Eight-fold Way', was in March 1963 invited to give a colloquium at Columbia University, New York, about the theory. One of the Columbia scientists, Robert Serber, later recounted how he had prepared for the talk by educating himself about the concepts of SU3. He quickly realized that families of one, eight, and ten emerge satisfyingly by combining the fundamental '3' in the combinations we just met, namely $3 \times 3 \times 3$ or $3 \times 3^*$. But why was it necessary to form combinations, and why these ones? It was all very well to explain the families of hadrons by combining two or three lots of '3' but the zoo of known particles contained no families of three, which seemed to be fundamental to the entire scheme. Serber made what is today realized as the correct use of the triplet: as three fundamental constituents that reproduce the SU3 families.

Prior to a talk by a visiting scientist, it was traditional for some members of the faculty to have lunch with the speaker. Serber was among the party. He put his idea to Gell-Mann, who replied: 'What are the electric charges of your particles?' Serber admitted he hadn't looked at this. Gell-Mann got out a pencil and on the paper table napkin figured out the answer in a couple of minutes. He announced that the charges would be $+2/3$ and $-1/3$ of a proton charge. As no such particles had ever been seen, and with such anomalous amounts of charge they could hardly have been missed, this was thought to be an appalling result. Nonetheless, during the colloquium Gell-Mann mentioned the idea and it was discussed also afterwards at the coffee gathering. Gell-Mann said that the existence of such a particle would be a 'strange quirk of nature', and quirk was jokingly transformed into 'quark', a word that had amused Gell-Mann when he read James Joyce's novel *Finnegan's Wake*.

In 1964, Gell-Mann wrote a paper in which he proposed the idea of quarks as constituents of hadrons, and he acknowledged Serber's input. Two other theorists also independently published very similar ideas at the same time: American George Zweig, and in Switzerland, Andre Petermann. In this picture, three varieties of quark—up, down, and strange—are the basic 3. The corresponding antiquarks—anti-up, anti-down, and anti-strange—form a 3*. The hadrons physically are then either baryons, formed from three quarks, or mesons formed from a quark and an antiquark.

Meanwhile, Serber had realized that although the fractional electric charges of quarks appeared peculiar, their 'magnetic moments' need not be. When a magnet interacts with a magnetic field it feels a mechanical torque, or turning force. The torque's magnitude is proportional to the intrinsic magnetism of the object and to its orientation with respect to the field. This measure is conventionally known as the object's magnetic moment. For example, the magnetic moment for a bar magnet is the product of the strength of either pole and the distance between them. In quantum theory the magnetic moment of a fundamental charged particle is proportional to that charge and to Planck's constant—the basic measure of quantum physics—and inversely proportional to the particle's mass. So, the electron, muon, and proton, which have the same magnitude of charge, will have magnetic moments proportional to the inverse of their masses. In other words, the electron will have a magnetic moment some 207 times larger than that of its more massive cousin, the muon, or 1836 times that of the proton.

While experiment confirms that the ratio of electron and muon magnetic moments scales perfectly with their relative masses, the same is not true in the case of the proton. Relative to the electron, a proton's magnetic moment is nearly three times larger than the simple theory would expect. Even more noticeable is the fact that the neutron, with no electric charge, has not only a measurable magnetic moment but a sizable magnetic presence, its magnetic moment being about 2/3 that of a proton and of opposite sign.

Serber's insight was that as magnetic moments of particles depend on the ratio of charge to mass, in a proton or neutron made of three quarks, each quark would have an effective mass 1/3 that of the parent, the 1/3 electrical charges and the 1/3 mass fractions cancelling out in the ratio, giving the quarks normal integral nuclear magnetic moments. Serber made a simple calculation and found that relative to the fundamental magnetic moment scale, the proton would have magnitude 3 and the neutron −2: values that are quite close to the measured ones, their ratio of −3/2 being within 2% of the empirical value.

Given the efficiency of the quark model in explaining the patterns and properties of hadrons, it may seem puzzling that the idea received such little interest at the time. Even Gell-Mann disparagingly dismissed 'concrete quarks' as being an idea 'for blockheads'. The manifest problem was that if quarks had masses smaller than the proton, these fractionally charged particles should have been found in experiments long ago. Their clear absence implied to Gell-Mann that either the idea was nonsense—though its many successes in collating and interpreting phenomena made that increasingly unlikely—or there was some as yet unrecognized further property that prevents fractional electric charges existing in isolation.

The magnetism of quarks

Until 1967, it was thought all particles were either neutral or carried electric charges that are integer multiples of those of the electron or proton. Unless one accepted that hadrons are not fundamental particles, there was still a puzzle of how a neutron with no electric charge could have magnetism. This was known more than 70 years ago and with hindsight, perhaps, was the first direct evidence that neutrons are not fundamental particles but have some deeper level of structure. Today we know this is correct. As the Earth has a magnetic field which arises from the swirling electric currents in its core, so does a neutron contain currents whose electric charges balance perfectly to zero but nonetheless give rise to an overall magnetic moment.

What is good for the neutron would also be good for the proton. The electric charges within the proton do not cancel out but add to give the proton its overall positive charge. As in the case of the neutron, their swirling motions will give rise to magnetic fields which on this occasion add up to a total magnetic moment. The question of course was: what is the nature of these internal constituents?

Although the quark model had already stumbled on the answer, this would only become clear in hindsight. Few took the idea of fractionally charged constituents seriously back in 1964. By that time, however, it was becoming evident that neutrons and protons have an internal structure and a measurable size, which inspired a quest to probe inside and see what they consist of. The scale of the challenge can be illustrated if we imagine the breadth of a hydrogen atom enlarged five trillion times, so that it is about 500 metres across. On this scale the size of a proton is about five centimetres. By analogy, the proton's extent within the atom is like the size of a hole on a golf course compared to the distance of the remote spot where a champion golfer is teeing off. While we can easily look inside that hole and retrieve the ball, to look inside a real proton and identify what lurks there requires light of some trillion times better resolution—shorter wavelength—than our eyes respond to.

By the late 1960s, a way to do this had been found. When an electron is deflected by an electromagnetic field it radiates light. If an electron can be speeded up to within one 100,000th part of the speed of light, or in energy terms to kinetic energies of tens of billions of electronvolts—that is, some 20,000 times the energy (mc^2) of an electron at rest—there is the possibility of it radiating light at wavelengths capable of resolving the innards of a proton. Indeed, if beams of such electrons are fired into a target of hydrogen, the electric fields surrounding the protons at the centre of the hydrogen's atoms can be sufficient to deflect the electrons, the resulting light having wavelength short enough to do so.

This was first done in the 1960s at SLAC, the Stanford Linear Accelerator Center, in California. At three kilometres in length this was the longest linear accelerator—or 'linac'—in the world. Electric fields first accelerated the electrons to an energy of 50 GeV (billions of electronvolts). Conceptually this is like Rutherford's experiments of half a century earlier, in which he fired massive alpha particles at atoms, discovered that the particles were scattered at much larger angles than anticipated, and deduced that the positive charge was concentrated on a massive entity—the atomic nucleus—rather than being smeared through the atom's extent. At SLAC, the beams consisted of lightweight electrons, their deflection being affected by any electrically charged entities within the proton, irrespective of how heavy or light those constituents might be. The discovery that electrons emerging from the accelerator were occasionally deflected by the hydrogen atoms through very large angles was a replay for a proton of what Rutherford had seen for the atom. Had the proton's charge been smeared throughout its extent, the electrons would have been less radically deflected; the large angle scattering was symptomatic of the proton's charge being carried instead by small particles. Today these particles are recognized as being quarks.

While Stanford was the first to discover the presence of these seeds, it was complementary experiments at CERN, using beams of neutrinos rather than electrons, that pinned down the constituents' properties. High-energy beams of neutrinos are much like electrons with their electric charge removed. Whereas electrons bounce off a target unchanged, deflected by the effects of the electromagnetic force, neutrinos pick up electric charge from the target proton thanks to the weak nuclear force and convert into electrons. The results of these complementary probes of the target can be combined to give more detailed information than comes from either experiment on its own.

The relative chance that a neutrino scatters due to the weak force is smaller than that of an electron via the electromagnetic force, and there are other differences in details related to the violation of mirror symmetry in the weak interaction, but these can all be considered when the results of the experiments are analysed. In effect, the chance of an electron bouncing from the target's constituents is proportional to the squares of the constituents' electric charges, whereas the chance for a neutrino to scatter is sensitive to the differences in their electric charges. Similar experiments using a target of deuterium, effectively heavy hydrogen, where the nucleus contains both a proton and a neutron, made it possible to study the innards of the neutron too. By comparing the sums of the squares against their differences, for both proton and neutron, it was possible to deduce the actual charges of their constituents. By keeping careful account of how the beams exchange energy with the target, and how this correlates with the amount of their deflection from the initial direction, it is possible to learn more about the constituents' properties. By 1973 the electrical innards of the proton and neutron had been resolved.

****

The electrically charged constituents of the proton or neutron target are surrounded by electric and magnetic fields. The resulting electromagnetic forces scatter the incoming beam of electrons. The electron is deflected through some angle and loses some energy in the process. The experiment records the result of millions of such collisions and builds up a distribution of how the probability depends both on the angle of deflection and the energy transferred.

From these data it is then possible to deduce the relative contribution of electric and magnetic fields produced by the target's constituents and learn about their nature. For example, a lump of static charge at rest is surrounded by an electric field to the exclusion of magnetic effects, whereas a spinning charge will give rise to a significant magnetic field. The results of the experiment at Stanford showed that the magnetic field is very important, implying that the constituents carry some spin. When the data were studied carefully, they showed that the source of the proton's charge is carried by constituents spinning with an angular momentum of 1/2, in units of Planck's constant. This is identical to the behaviour of the electron itself. This result, which first appeared in 1969, showed that the proton is composed of more fundamental constituents—quarks—which are spinning at the same rate as does the fundamental electron.

A proton has spin 1/2, and for it to be made of constituents that themselves have spin 1/2 implies that there must be an odd number of these units. Similar results were found when beams of neutrinos scattered from protons. As neutrinos also transfer electric charge to the target this required that there are at least two varieties of constituent, one of which picks up the charge transferred from the beam and is in the process converted to the other variety. The difference in the charge of these two constituents must be the same as that of a neutron and proton.

The probability for an electron to scatter is proportional to the square of the charge of the constituent that it hits. The data showed that the sum of the squares of these charges is 5/9 relative to the charge squared of a proton. When all the data were analysed, they showed that the simplest construction of a proton and neutron is for them to be composed of three constituents with charges that are either +2/3 or −1/3 relative to the single positive charge of a proton. These are known whimsically as *up* and *down* quarks, denoted by the symbols *u* and *d*, respectively. The difference in the magnitude of their respective charges is indeed the same as that of proton and neutron: $1 = 2/3 - (-1/3)$, the sum of the squares being $5/9 = (2/3)^2 + (-1/3)^2$.

Isolated quarks have never been found. Nor do they occur in pairs or quartets. They cluster in triplets. The proton is then most simply configured as *uud*; the neutron is *ddu*. Three up quarks give a total charge twice that of a proton, whereas three down quarks give a unit of negative charge. These have been seen to occur in short-lived particles known as Delta resonances.

The difference in the magnitudes of up and down quarks' electric charges precisely equals the 1.6×10^{-19} coulombs of the quantum of electric charge as carried by an electron. The quarks spin at the same rate as the electron, and their magnetic moments are in proportion to their electric charges, again in direct proportion to the magnetic moment of the electron. The magnetic moment of the up quark is empirically twice that of the down, and of opposite orientation, in accord to their relative electric charges being 2/3 and –1/3, and when the relative masses of quarks and electron are taken into account, the quarks' magnetic moments are indeed the relevant one-third fractions of an electron's. Everything is consistent with quarks and electrons being fundamental particles, their response to the electromagnetic force being controlled by the amount of their respective electric charges.

****

Recall what had excited Robert Serber when he raised with Gell-Mann the idea that became 'quarks'. Although the quarks' electric charges are 1/3 fractions of the proton's total charge, their magnetic moments are also inversely proportional to their masses. In the simplest picture, where a proton's mass is also shared among the trio of quarks, each quark's fractional mass counterbalances its fractional charge such that the quarks' magnetic moments are then integers relative to those of the proton. Quarks have unusual electric charges, but their magnetic moments are more rational.

Having identified that proton and neutron are made of quarks, the mystery of the neutron's magnetism can be explained. Its vanishing electric charge is easy to understand, as the sum of the quarks' charges in the neutron—*ddu* in quark content—is 0 thanks to the –1/3 charges of the two *d* quarks countering the +2/3 of the lone *u* quark. Their magnetic moments, however, do not mutually cancel one another. The reason is rooted in quantum theory, but its essence is easy to illustrate.

A quark has a magnetic moment because it spins, quantum mechanics allowing the resulting magnetic moment to be oriented in one of two directions, either parallel or anti-aligned relative to the lines of force in an applied magnetic field. We can imagine this as the quark having either its north or south magnetic pole pointing in the direction of the field.

All fundamental particles are classified as either bosons or fermions. Bosons are particles which have integer or zero amounts of spin. Bosons can congregate collectively in unlimited numbers sharing the same quantum states; photons, for example, which have one unit of spin, can collectively form laser beams of arbitrary intensity. Fermions, which we have met before, have 1/2 integer spins, like the electron, neutrino, or quarks, and by contrast are exclusive. If one fermion already occupies a quantum state, no other fermion can be accommodated there, and instead must find a vacant situation elsewhere. This exclusion principle has great scientific provenance, especially in chemistry where, when applied to electrons in atoms, it explains the chemical periodicity of the table of elements. It is this exclusion principle that is fundamentally responsible for every eighth of the lighter element of the periodic table being an inert gas, namely helium at number 2, neon at 10, argon at 18, and so on. By contrast, their neighbours, with one electron extra or fewer than the above numbers, are chemically very active.

The principle is also key to understanding how the magnetic contributions of the quarks within a neutron or proton combine. This is in effect what Serber calculated back in 1963, following

Table 1 Neutron and proton magnetic moments in the quark model. The two identical flavours are each weighted by a factor of 2, and the third quark by −1. Reasons for these weighting factors are given in the text. The sums then give the relative scales of magnetic moments of the neutron (−2) and proton (+3).

Neutron	Proton
d: 2 × (−1/3)	u: 2 × (2/3)
d: 2 × (−1/3)	u: 2 × (2/3)
u: −1 × (2/3)	d: −1 × (−1/3)
Total: −2	+3

his and Gell-Mann's epiphany. In the neutron, the quantum constraints on the orientations of the quarks' spins tend to align the spins of the two *d* quarks at the expense of the lone *u* quark whose spin—and magnetic moment—is flipped in the opposite direction. The quantum accounting is intricate, the results being that the two like quarks (*d* in this case) get over-weighted by a factor of two and the lone one (the *u*) gives a contribution oriented in the opposite direction. So, in the case of the neutron, two *d* quarks, each with charge −1/3, contribute a total of −4/3 to the overall magnetic moment. Meanwhile, the up quark with charge +2/3, is flipped, pulling the previous value of −4/3 down to −6/3, or −2 (see Table 1).

The analogous calculation for the proton replaces every *d* by *u* and every *u* by *d*. In this case, the two *u* quarks with charge 2/3 are doubly weighted giving a total of 8/3, and the lone *d* quark with charge −1/3 has its magnetic moment flipped, to give a contribution of +1/3. Adding this quantity to the previous 8/3 give a total of 9/3, or 3. The ratio of the neutron and proton magnetic moments is then –2/3, in remarkable agreement with what has been experimentally measured for these two composite particles. So not only is the mystery of the electrically neutral neutron having a finite magnetic moment resolved, but the magnitude is also understood once we have realized that both neutron and proton are made of three spin-1/2 quarks.

6

A QUARK'S COLOURFUL WORLD

While the weak force can change one variety of atomic nucleus into another, and the attraction of opposites entraps the negatively charged electrons around the positively charged nucleus, it is the strong nuclear force that enables large numbers of positively charged protons—26 in the case of iron—to survive in close contact even while electrostatic repulsion is forcing those like charges apart. Empirically it acts on protons and neutrons only when they are in close contact, which is such a short range that we are unaware of this force in the world at large.

Today we know that this strong force arises from the quarks that build hadrons. When protons or neutrons touch or overlap, one or more quarks can swap from one particle to its neighbour. In quantum mechanics this gives rise to an attractive 'exchange' force. In this case, the attraction of the nuclear constituents is a consequence of the forces that glue quarks to one another in trios to make neutrons and protons. At the fundamental level of quarks, it transpires that there is a remarkable similarity between this force and the electromagnetic force.

Electric charge is the source of electric and magnetic forces. The strong force is a result of another form of charge carried by quarks but not by leptons. This charge is colloquially referred to as colour, which is just a name as it has no relation to real colour; the reason for the choice will become clear later. Leptons, like the electron or neutrino, do not carry colour charge and so are blind to the strong force; protons and neutrons which contain quarks and consequently contain colour charge do feel the strong force. It is when we focus on the quarks themselves that profound similarities between the behaviour of colour charge and electric charge become apparent.

First, recall the rules of *electrostatics*: like charges repel, opposite charges attract. So it is that a negatively charged electron is attracted by the positively charged nucleus. An electron and its antiparticle, the positively charged positron, can also attract one another for so long as they do not meet and mutually annihilate. Very similar rules apply in the case of colour charge surrounded by a 'chromostatic' field. Quarks carry the positive colour charge by convention, antiquarks carrying negative colour charge. The attraction of opposites causes a quark and antiquark to have a mutual affinity for as long as they do not meet and mutually annihilate. Experiments in cosmic rays and at particle accelerators have revealed countless examples of mesons, such as pions, kaons, and others, most simply made of a quark bound to an antiquark.

The key difference between electric charge and colour charge is that the latter comes in three distinct varieties. It is to distinguish them that they are traditionally referred to as colours, red (R), blue (B), and green (G). As is the case for electrostatics, the rule of chromostatics is that like colours repel. So, two quarks each carrying a positive red charge will mutually repel one another, as will two greens or two blues. A quark with a red charge and one with a blue have the possibility of attracting. The full calculation of whether they attract or repel each other depends upon profound details of quantum theory which are not relevant here. The important result is that there is a 50% chance that an unlike pair will attract one another, and a 50% chance that they will repel. So, we might say that two unlike charges attract but with effectively half the strength compared to that of two like charges repelling one another.

As there are three different coloured charges, it is possible for three quarks to attract one another so long as none of them carry the same colour charge. Three quarks with colour charges red, blue, and green can happily coexist thanks to their mutual attraction. It is now obvious why protons and neutrons are most simply formed out of threes, for consider what happens if a fourth quark appears on the scene. This interloper will itself carry a colour charge; let's suppose it is red. Being different to the blue or the green colours in the neighbouring trio it can be attracted to each one of them, though with only 1/2 as much strength as the repulsive force exerted on it by the like-coloured quark, the red one (Figure 8). As two halves make a whole, so will the two half-hearted attractions perfectly counterbalance the like-charge repulsion. The result is to leave the trio overall effectively neutral as regards colour attractions and repulsions. In the jargon we say the colour forces have 'saturated' when three different colour charges combine this way.

The three colour charges therefore explain why the quarks like to cluster in threes. This has not resolved the puzzle of electrical neutrality, however, though it might have moved us nearer to finding a solution if we could construct a viable theory that

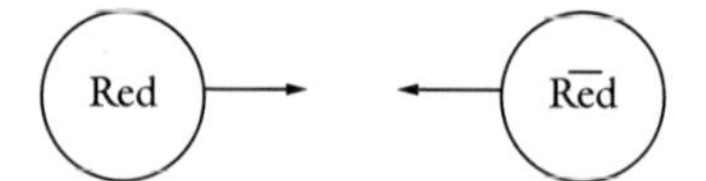

Colour and opposite colour attract. Thus is formed a **meson** $q\bar{q}$.

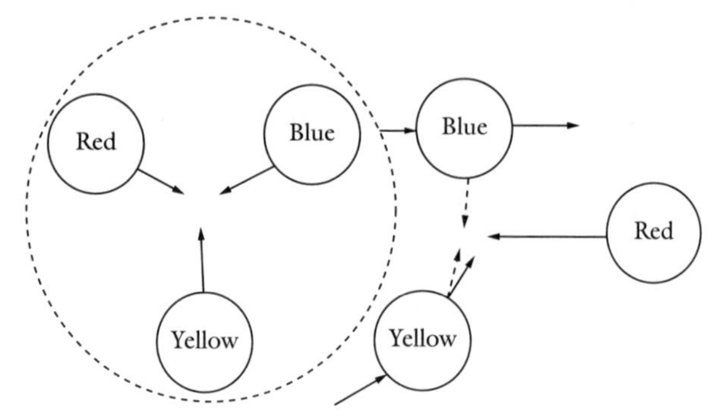

Three different colours attract. Nearby like colours are repelled. Colourless clusters of three different coloured quarks form. Hence **baryons** $q_R q_B q_Y$.

Figure 8 Colour charges at work (upper part of figure). Colour and opposite colour attract. Thus is formed a meson (lower part of figure). The attraction of three different colour charges creates a state that is overall 'colour-blind'. A fourth quark must carry the same colour as one already present. The repulsion by that identical colour balances the two 'half-strength' attractions by the two different colours.

links colour charge and electric charge. The electron, with the canonical amount of electric charge, does not feel the strong force because it has no colour charge. The quarks, by contrast, feel the strong force because they carry colour charge, and in three varieties. Meanwhile their electric charges have fractionated into amounts that are 1/3 and 2/3 relative to the charges of the electron and proton. Three colours cluster quarks in trios whose individual members have 1/3 fractional electric charges: a tantalizing piece of numerology—that $3 \times 1/3 = 1$—leads to the electrical balance.

If this approach is on the right lines, it would imply that not only are quarks and leptons intimately related, for which we already have found some evidence, but that the strong and electromagnetic forces, manifested by the three colour charges and single electric charge, respectively, are also fundamentally unified. We have already seen hints that this may be true, as in the commonality of the rules of attraction and repulsion for both electric and colour charges. The more closely we investigate the properties of electromagnetic and colour forces, the more similar they become.

Magnetic colour

The spin of the electrically charged electron and quarks gives each one of them a magnetic moment. This spin also acts on the quarks' colour charge, giving them a 'chromo-magnetic' moment. This colour magnetism behaves in a completely analogous way to its electro-magnetic cousin and gives a corresponding set of phenomena.

The electron's magnetism is manifested in the spectral lines emitted by atoms when they are heated. Add energy to an atom and one or more of its electrons will be raised from low-energy to higher-energy quantum levels. When they then fall back to lower-energy states, they emit the excess energy in the form of electromagnetic radiation. The consequent discrete set of spectral lines reveals the existence of distinct energy levels in the atom. Indeed, it was the discovery of spectra back in the nineteenth century that, with hindsight, gave the first indications that atoms have an internal structure.

In 1896, the Dutch physicist Pieter Zeeman discovered that in the presence of a magnetic field, an atomic element's spectral lines can split into individual components, typically varying from two to a handful. The separation between the individual lines is proportional to the strength of the magnetic field. This shows that the splitting is linked to the magnetic moment of the atom, or at least of the constituents responsible for the spectral lines. The splitting is known as 'fine structure' of the spectral lines.

In quantum mechanics, the number of orientations of a magnetic particle, atom, or molecule in a magnetic field is constrained. To determine the amount, start with the magnitude of its total angular momentum when measured in the traditional units proportional to Planck's constant, double it, and add one. For example, spin angular momentum of 1/2 will give two lines, an amount of 1 gives three, and so on. Each of these orientations will have a different energy. When the magnetic moment is aligned with the field, the energy is lowest; when pointing head-on into the field, the energy is maximal, with the other orientations having their energies equally stepped—'quantized'—from the maximum to the minimum value. From the number of these lines, the atomic spectroscopist can determine the total angular momentum of the object.

The simplest example is that of hydrogen, a system of two particles, comprising a single electron and proton. When the electron is in the lowest-energy state, it has no overall rotary momentum around the central proton. Even when there is no external magnetic field, however, the electron will feel a magnetic force due to the magnetic moment of the proton at the atom's centre. In quantum theory, the electron having spin 1/2 can take up two

(1 added to twice 1/2) possible orientations relative to the spin of the proton. The magnetic moment of the electron can align in the same direction as the proton's magnetic moment, or opposite to it. These two configurations have slightly different energies.

Imagine that the electron in the higher energy of these two configurations reduces its energy by flipping its spin, and hence its magnetic interaction. It ends up in the lower-energy state, the difference in these energies being emitted as electromagnetic radiation. The wavelength of this radiation is about 21 centimetres, some million times more than visible light, which makes it part of the radio spectrum. Radio astronomers have tuned their detectors to pick up this radiation (its frequency is 1420 MHz, some 1.4 billion cycles per second, which is in the vicinity of some UHF television signals). With their receivers tuned to this frequency, they can use this natural phenomenon to detect the presence of hydrogen in the cosmos.

In the jargon of atomic physics, this is known as 'hyperfine splitting'. It is proportional to the magnetic moments of the electron and the proton. Whereas fine splitting in atoms involves just the magnetic interaction of the atom's electrons alone, and is in inverse proportion to an electron's mass, the hyperfine splitting involves the magnetic moment of the proton as well, which is triflingly small by comparison due to its extra heft. The magnetic splitting of the spectral lines is thus much smaller, hence 'hyper'-fine. The higher-energy configuration is when the electron and proton spins are oriented parallel to one another, the lower-energy state being when they are spinning in opposite directions.

So much for these effects when electric charges and magnetic moments are involved. What is the analogy when electric charge is replaced by colour charge?

The simplest example is to compare the hydrogen atom, made of two oppositely charged constituents, with mesons made of two oppositely coloured entities: a quark and an antiquark. The quark and antiquark, of course, also carry electric charges and have conventional magnetic moments, just as do the electron and proton. So, the energy of the quark and antiquark in their lowest quantum state will be subject to an electromagnetic hyperfine spitting, analogous to what we have already seen in the hydrogen atom example. However, the magnitude of this electromagnetic effect is negligible compared to its colour magnetic analogue. It is the 'chromo-magnetic' effects that are most noticeable.

When the electron and proton are aligned in a hydrogen atom, their energy is greater than when they are spinning in opposite senses by just a fraction of an electronvolt. When a quark and antiquark's spins are aligned, here too the energy is larger than when they spin in an opposite sense. The difference with the hydrogen example is one of magnitude. Indeed, in the case of colour charge, the two different spin alignments of quarks in mesons result in energies so different that they are named as different mesons. There are two mesons whose masses—energies at rest—fit with this phenomenon. The lower energy state is the pion, and the higher is known as the rho-meson; these are separated by about 600 million electronvolts. Qualitatively the spin-dependent effects in the world of the coloured quarks and of electrically charged electrons and protons are similar. Quantitatively, however, they are very different.

The reason is twofold. First, the intrinsic strength of the interaction between colour magnetic moments, linked as they are to the source of the strong force, is much larger than that between the electromagnetic equivalent. Second, the breadth of a meson is only about 1/100,000 that of the hydrogen atom, or in volume a mere thousandth of a trillionth. This means that the magnetic interaction has only to stretch a trifling fraction of the extent that it could in the hydrogen atom case and as a result is correspondingly more likely. It is the combination of these two effects that explains the billionfold difference in energy scales in the two cases. Once that is allowed for, everything else about the two situations is remarkably similar.

Quantum chromodynamics

We have seen that colour charge appears to be like electric charge split three ways, and now it is apparent that colour magnetism and electromagnetism follow the same fundamental path. It hardly requires a great leap of genius to suspect we have stumbled upon a profound truth that links the strong force at some fundamental level with the well-known electromagnetic force.

Recall that the marriage of electric charge with relativity and quantum field theory led to the empirically successful theory known as quantum electrodynamics. The essence there, as in all quantum field theory, is that a particle will be accelerated or decelerated when it experiences a force, and that in the case of electric charge this is the electromagnetic force. By analogy, if colour charge is the source of a force, then we may expect that when colour charge is combined with relativity and quantum theory a similar mathematical description of this force will emerge. The resulting theoretical construct is known as quantum chromodynamics (QCD).

Recall that in quantum field theory, a force transfers energy and momentum from one particle to another by the intermediate action of a particle—the carrier of the force. In the case of quantum electrodynamics, the carrier is the photon, the massless corpuscle of light. The mathematical structure of QCD leads to a completely analogous situation in the case of colour charges. The force is now transmitted by massless particles known whimsically as 'gluons'. What photons are to the electromagnetic force and quantum electrodynamics, so gluons are to the colour force and QCD. It is the action of gluons flitting between the threefold-coloured quarks that 'glues' them together in threes, forming the clusters that we recognize as neutrons and protons.

In quantum electrodynamics, photons are radiated when an electron changes its state of motion. That is how beams of electrons were scattered by electromagnetic forces in and around a target proton in the experiments that first revealed its constituent quarks, which in turn led to the breakthrough in finding unity between the electromagnetic and strong forces inside this 'femto-universe', as the Stanford theorist James Bjorken referred to the restricted Universe no more than a femtometre (10^{-15} m) across that is home to quarks within a proton or neutron. Analogously, in QCD the sudden enforced change in motion of a colour charge radiates gluons.

That QCD is empirically true has been demonstrated in many high-energy physics experiments. The first clear example came back in the 1970s when electrons were annihilated by head-on collisions with their antiparticle counterparts, positrons. Because the electron and positron have the same mass but equal and opposite amounts of charge, a magnetic field can swing beams of them along the same arc but in opposite directions. This is the principle that underpinned the advent of electron–positron colliders in the latter half of the twentieth century. The positrons had to be contained in an evacuated tube for their preservation. The goal was to make two counterrotating beams of positrons and electrons meet head-on. This is where the full power of the high-energy particle and antiparticle can be brought to bear.

The electron and positron annihilation destroys their material presence. Their opposite electrical charges having cancelled out, no electric charge remains, but like the smile of the Cheshire cat, their energy remains, manifested in a flash of light. Most likely they will have converted into a single photon. This is not a photon of real light with mass zero, however. In the head-on collision of an electron and positron, travelling at the same high speeds but in opposite directions, their total energy adds up to twice their individual energy while their momenta are equal and opposite, and cancel to zero. This results in a photon at rest, with a finite amount of energy. Such a physical imbalance arises thanks to quantum mechanics, courtesy of the uncertainty principle which allows energy conservation to be overridden for very brief timescales. Consequently, this situation can only last for the merest fraction of a second, after which time this 'virtual photon' converts back into counterbalancing particles and antiparticles.

These products might be simply an electron and a positron, replicating what created this situation in the first place. More likely is that new forms of matter and antimatter, such as a quark and antiquark, emerge from this miniature fireball. The transient electromagnetic flash had no charge and no colour so neither can the products overall. Consequently, their electric and colour charges both mutually cancel.

When a quark and antiquark are produced this way, they fly off in opposite directions carrying colour charges with them. The colour field between them grows as they fly apart, the energy of their motion being transferred to this field. As soon as this energy is large enough to create a further quark and antiquark, it will do so. The result is that before the original quark and antiquark have travelled a femtometre, 10^{-15} m, two oppositely directed 'jets' of particles made of quarks and antiquarks will have been created, like some form of quantum zipper. The two jets of particles and antiparticles fly apart from one another, their direction revealing the motion of the quark and antiquark that spawned them. The orientation of the axis of these jets relative to the direction of the original electron and positron empirically agrees with what theory would expect if the quarks had spin 1/2, giving further proof of this fundamental property.

The sudden appearance of electric charges, and their consequential acceleration from rest, can radiate one or more photons. The relative angle between two jets of quark and antiquark and the direction of the photon is further empirical confirmation that the photon is radiated by spin-1/2 quarks and antiquarks.

Now for the culmination of this piece of the story. As quarks and antiquarks carry not just electric charge but also colour charge, not only do they radiate photons but also gluons. QCD predicts how the jets spawned by the quark, antiquark, and gluon are distributed around the compass. As the only difference between the former case with photons and this with gluons is the replacement of electric charge by colour charge, everything about this distribution should be the same. The results of experiments agree perfectly with this prediction.

This is but one example among many, but it is one of the easiest and cleanest demonstrating not only that gluons act like coloured photons, but that the theory of QCD is the correct mathematical description of the behaviour of quarks and gluons within the femto-universe.

Trapped in the femto-universe

These tempting similarities between electric and colour charges, leading to quantum electrodynamics and QCD, beg the question as to whether this is all an accident or whether it indicates deeper connections that might ultimately resolve the puzzle of the electrical neutrality of atoms. These similarities are only noticed when you probe nature deep within the femto-universe. At larger distances, even at atomic or nuclear dimensions, the electromagnetic and strong forces appear very different.

The very name 'strong force' illustrates an obvious difference with electromagnetic forces. When acting on neutrons and protons within an atomic nucleus, the strong force is at least 100 times more powerful than the electromagnetic force; where it any less it would hardly be able to overcome their electrostatic repulsion. At least, that is how its effects appear when at work on a distance scale of 10^{-15} metres, the distance that neutrons and protons must encroach to bind strongly in complex nuclei.

The strength of the force at these distances illustrates how quarks cluster to form the proton itself. It also explains why it is impossible to pull individual quarks out of one of these colourful clusters. Instead, groups of three appear to be permanently confined inside the femto-universe. If the quarks are separated by distances of this order, the potential energy in the colour field between them becomes overly large. It can reduce to greater stability by converting into a quark and antiquark. This leads to the decay of the original cluster into two particles, one of which would be a meson made of one of the original quarks and the new antiquark spawned from the colour field. Instead of isolated quarks appearing in these circumstances, therefore, mesons typically are spawned.

This is very different from the case of the electromagnetic force, where it is relatively easy to ionize atoms. For example, temperatures of a few thousand degrees are sufficient to liberate one or more electrons from an atom, leaving it with a net positive charge. This is known as a 'positive ion'. Similarly, electrons can be added to atoms, making negative ions. It is the electrical attraction between positive and negative charged ions that is key to producing some chemical compounds such as sodium chloride—common salt.

A reason why ionization happens for electric charge but not in the colour analogue lies in their different behaviours as a function of distance. When electric charges become further apart, the force between them dies off as per the inverse square of the distance between them. For colour charges the behaviour of the force is nearer to a constant, the potential energy within the colour field growing linearly with the distance of separation.

The behaviours of electric and colour forces as a function of the distance are so dissimilar that it might seem like a sleight of hand to be claiming these two forces are fundamentally alike. The explanation is that although the effects are very different when studied at distances larger than a femtometre—indeed there is no obvious sign of the strong force at all in the world around us beyond our deduction that it must be there to bind the nucleus—for distances much smaller than this, the colour and electromagnetic forces are indeed increasingly alike.

First, recall the reason for the inverse square law of attraction in the case of electrical charges. This arises because the electrical field surrounding the charge has no preference for direction and spreads uniformly throughout all of space. In effect the photons travel outwards independently and, having no electric charge themselves, they feel no mutual attraction. If we imagine a series of shells surrounding the source through which the photons pass, the surface area of the shells grows like the square of their radius. This means that the intensity of photons will die away as per the square of the distance from the source. This is in effect the source of the inverse square law for the resulting electromagnetic force.

If the gluons of the colour charge were themselves blind to the colour force, then here too there would be an inverse square law as the gluons propagated away independent of one another. However, the fact that there are three varieties rather than one variety of colour charge gives the opportunity for the gluons themselves to carry colour. Gluons in effect carry colours in pair-wise combinations, an individual gluon being able to remove any of three colours from a quark and then 'repaint' it with any of the three colours. For example, when a red-coloured quark converts to a blue quark in the process of radiating the gluon, that gluon must carry off the red colour and bring in the blue. As gluons themselves carry colour, they can mutually attract one another by the same colour force that spawned them. The result is that as they propagate through space, they are not independent of one another. The actual behaviour of their propagation is very complicated to compute, but the effect empirically is that the energy within the colour field grows in proportion to the distance. The difference in the long-range behaviour of the strong colour force as against the relatively feeble electromagnetic force is therefore simply due to the replacement of one by three in the numerical variety of the charged sources. At long distances nature has disguised what is revealed deep in the femto-universe to be a profound harmony.

7

JANUS-FACED QUARKS

The quark layer of reality, where QCD acts like QED 'writ three times', is also the theatre where nature uses a similar scheme 'writ twice'. That beta decay can convert one element in the periodic table into either of its immediate neighbours has been known for over a century. Now we recognize this to be the result of a fundamental dynamics where one variety of quark converts into another as the result of the so-called weak force. The mechanism appears to be once again driven by a charge; not a loner like electric charge or a trio like colour, but a charge having a duality.

The charge that gives rise to the weak force is known by the name of 'flavour'. Just as colour charge comes in three varieties so does flavour come in two, hence the allusion to QED 'writ twice'. The down and up quarks are examples of quarks that are Janus-faced, carrying one or other of two different flavours—which we label *up* and *down*. The most familiar effect of the weak force to which the flavour charge gives rise is that it can switch one face into the other.

The names 'up' and 'down' for these siblings reflect how the mathematical accounting scheme for two flavours has been taken over from that for the magnetism of spinning particles. In the case of magnetism, a spin-1/2 particle's magnetic moment can be aligned either parallel—'up'—or counter—'down'—to the direction of the magnetic field. Relative to some agreed unit amount of energy, the up orientation is traditionally referred to as having gained magnetic energy proportional to +1/2 while the down orientation has the energy reduced proportional to −1/2. By analogy in the case of flavour, the up configuration traditionally is said to have flavour charge +1/2 and the down configuration has −1/2, scaled relative to some agreed quantum of flavour charge.

If a down quark absorbs one positive unit of flavour it will convert to an up quark. Conversely if an up quark absorbs one negative unit it will convert to a down quark. The same is true for emission rather than absorption. An up quark emitting a positive unit flips into a down, while a down emitting one negative unit becomes an up.

The entities that carry these units of flavour charge are known as W bosons. When the theory of flavour charge is made consistent with relativity and quantum field theory, by analogy with what was done in the case of electric charge, the W bosons emerge as analogues of the photon. They are superficially like electrically charged photons, except that empirically they are massive, whereas the photon has no mass. (Reasons for this will come later in the section on the Higgs boson.) The difference in flavours of one unit in this case also matches the electric charge of the W. For example, when a positively charged W flips a down-flavoured quark to an up, both the flavour and electric charges of the quark change by one unit (Figure 9).

For the W boson, electric charge and flavour charge match, the unit positive electrical charge also being a positive unit of flavour, whereas the unit negative electrically charged W has a negative unit of flavour. Electric charge is not the same as flavour charge in general, however. For example, the up and down quarks have electric charges in amounts that are fractions, +2/3 and −1/3, respectively, of the unit positive charge, not +1/2 or −1/2. Nor is the photon truly the flavour-neutral partner to the W^+ and W^-. The fundamental particle playing that role is more familiar as the Z^0, whimsically known as 'heavy light', reflecting that both the Z and the W bosons are very massive, their energies at rest some 80–90 GeV, nearer to the mass of a silver atom than to a massless photon.

In quantum theory, the probability of something happening is determined by squaring an algebraic quantity known as the

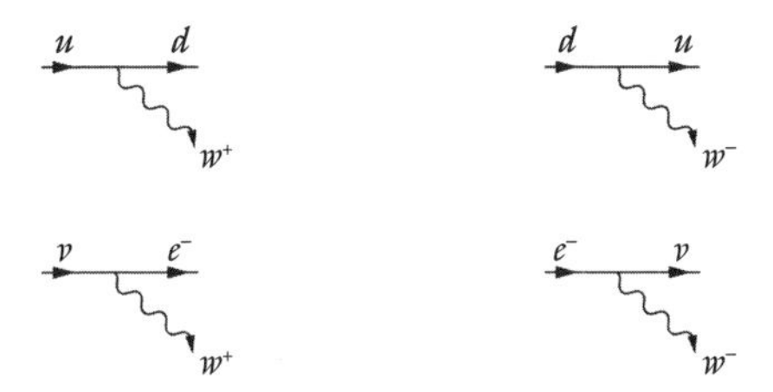

Figure 9 Flavour transmutation for quarks and leptons. The emission of a W boson carrying one unit of positive or negative charge changes the up or down flavour of a quark (a). The analogous processes for neutrino and electron (b). If W bosons were the only probes available to us, quarks and leptons could not be distinguished.

quantum amplitude. Probability is always positive or zero, so a quantum amplitude can be either positive or negative. For example, the chance of an electron being affected by an electric field is the same as for its positive analogue, the positron, but the quantum amplitudes for the two processes have opposite signs, being otherwise identical.

Whereas the quantum amplitude for a photon to interact with a particle is in proportion to that particle's electric charge, that of the Z^0 is proportional to the particle's flavour charge. So, its interaction with a down quark is the same magnitude but of opposite sign to that with an up quark, these quarks having charges on the flavour scale of $+1/2$ and $-1/2$, respectively. The physical meaning of these numbers is that the strength (technically the quantum amplitude) with which a Z^0 couples to a quantum of flavour charge is weighted by a factor of $1/2$, either positive or negative.

What is good for quarks is good for the leptons too. The neutrino and electron are flavour charged $+1/2$ and $-1/2$, respectively, with transitions between them involving the emission or absorption of a W^+ or W^-. Their interactions with the Z^0 are also proportional to $+1/2$ and $-1/2$ on the flavour scale. The quantum field theory describing the interactions of flavoured quarks and leptons with the carriers of the resulting weak force, the W and Z bosons, is known as quantum flavourdynamics (QFD).

Quarks in the mirror are like leptons

The property of parity violation played a key role in the experiments that used neutrinos to explore the innards of the proton. As in the parity violation experiment described in chapter 4, protons accelerated to high energy are smashed into a target, which produces pions. These fast-moving electrically charged pions are then focused into a beam by electric and magnetic fields, and their decays in turn produce beams of muons and muon-neutrinos. The parity violation experiment focused on the muons, but if muons are kicked away by application of magnetic fields, a linear intense beam of high-energy neutrinos remains.

Whereas low-energy neutrinos, such as are produced in beta decay or by nuclear reactions in the Sun, can pass through the Earth as if it is transparent, high-energy neutrinos are much more likely to interact. By placing a dense target of hydrogen in line with the neutrino beam, there is a reasonable chance that the neutrinos will burrow into the target's protons and be scattered. In so doing they pick up electric charge and the outgoing charged particle (most likely a muon in this experiment) can be detected. As the direction and energy of the incident neutrinos are known, the amount of energy and momentum exchanged with the target can be inferred. In addition to confirming that the proton's constituents have spin 1/2, as the electron scattering experiments had already independently demonstrated, the neutrino experiments revealed that these constituent quarks violate parity maximally.

Not only do the down and up quarks differ in their electric charge by 1.6×10^{-19} coulombs, identical to that between an electron and a neutrino, but the maximal parity violation favouring left-handed leptons is also matched precisely by the preference for left-handed down and up quarks. Here we see a further profound similarity between the properties of fundamental leptons and the basic quarks. Indeed, if the only means to access leptons and quarks was by the W bosons of the weak force, there would be no way to distinguish the pair of down and up quarks from the analogous lepton pair of electron and neutrino.

The result of all this is that the basic constituents of the proton and neutron appear to have almost identical properties to those of the leptons as far as the electromagnetic and weak forces are concerned. The quarks respond to the weak force the same as do the leptons, and to the electromagnetic force also once one takes account of the absolute magnitudes of their various electric charges. Within the scope of electromagnetic and weak interactions, these assorted amounts of electric charge are effectively the only real difference between these two classes of particle.

The tantalizing conclusion is that while the absolute difference in the electric charges of the pairs of leptons and of quarks is the same, the absolute magnitude of quark charges is on average based on a quantum that is 1/3 the size of that of the leptons. That the average charge of two up quarks and one down quark is 1/3, and the enigmatic coincidence that the quarks cluster not in random odd numbers to make the proton or neutron but in a trio, not one, two, or four, constitute the reason for the perfect electrical balance between the positively charged proton and the negatively charged electron.

We don't know why there is such a fundamental connection between leptons and quarks. Certainly, an electron is not made of quarks. According to the best experiments today, the behaviour of electrons is in precise agreement with QED and shows no sign of any more fundamental structure. Electrons and quarks seem to be basic elements of the familiar material world, at least on the level of resolution currently accessible—some 10^{-18} m. The electrical neutrality of atoms arises from a coincidence of threes between quarks and electrons, particles that have no known mutual relationship. If you are happy to buy that as coincidence, then you have answered the puzzle of the atom's neutrality. But why do we not see this one-third fractional quantum directly in our experiments? What is special about three that causes a trio of quarks to cluster this way, causing this remarkable balance of electric charge to happen?

The 1, 2, 3 of charges

In none of these cases—electric, colour, flavour—can I tell you what these charges actually are, but I can tell you how they behave and how they reveal themselves. And as we have seen already, the way they behave is remarkably similar one to the other.

When we say that a particle carries a charge –1 or +2/3 this is its magnitude relative to some standard unit amount of charge. In the case of electric charge this is the amount carried by a proton, or the negative of that carried by an electron. The absolute magnitude of this unit is 1.6×10^{-19} coulombs, so 1 coulomb is the amount of charge carried by some 6×10^{18} electrons. As we cannot realistically accumulate exactly the required millions of trillions of electrons together, this is not very helpful as a definition. To give an idea of magnitudes, if you rub materials to induce electrostatic attraction, you will have transferred a few micro-coulombs, that is, millionths of a coulomb. A bolt of lightning, as we have seen, transfers typically 15 coulombs, though devastating ones up to 350 coulombs have been recorded. A smartphone battery can hold up to 10,000 coulombs of charge. All of this still raises the question: what is a coulomb?

An attempt to find the answer on the Internet can lead you into a series of circular definitions. Electric current is electric charge in motion and the amount of current is measured in terms of amperes or 'amps'. This leads to a definition of 1 coulomb as 'the amount of electric charge delivered by a current of 1 amp in one second'. While true, this leaves hanging the definition of an amp. Search the Internet once more and you can complete the circle with: 'One amp is the flow of 1 coulomb per second.'

There are other ways of defining these quantities that are unambiguous, though harder to visualize. As much of our interest in charges is focused on their role as the source of forces, the coulomb and the amp can be defined in terms of the amount of force exerted by one charged object on another at some agreed distance away.

When an electric current passes along a wire, a magnetic field is generated. When currents flow along two wires, the one will exert a magnetic force on the other. Force is measured in newtons, where one newton is the amount of force that would give a mass of one kilogram an acceleration of one metre per second per second. An ampere can then be defined as the current that needs to pass through two very long wires, placed 1 metre apart, that would produce a magnetic force of 2×10^{-7} newtons per metre. The numbers here may appear strange but at least the principle should be clear. Having defined one amp, then in one second it will deliver one coulomb of charge.

While this definition is reasonably practical to measure, its logic is rather indirect if what one is really interested in is the meaning of coulomb. For our purposes, a coulomb of charge is the source of an electric field which will give measurable force, in newtons, on another electric charge. The force between the two charges follows Coulomb's law—that the strength of the force falls as per the square of the distance between the two charges. If two tiny particles each with 1 coulomb of electric charge are about 1/3 mm apart, the electrostatic force that one gives to the other will be 1 newton. This is easy to comprehend theoretically, but hard to do in practice, not least as a force of one newton would speed one kilogram at one metre per second per second, so a tiny particle, with mass a mere fraction of this, would fly off immediately unless somehow constrained.

By whatever means one decides to define the terms, the quantities are ultimately specified in terms of the amount of force they induce on other charged particles. 'Charge' is the most fundamental source of the phenomena.

Having defined the strength of the electromagnetic force in terms of numerical multiples or fractions of coulombs, the analogous measures of charge for colour and flavour, the sources of the strong and weak forces, respectively, follow directly. Although years of familiarity with electromagnetic theory have led me to accept that electric charge is measured in coulombs, I still have an unsatisfied urge to know what a coulomb is. For example, what are its dimensions in terms of familiar measurables like length L, time T, or mass M?

In most applications, the magnitude of the charge appears squared. For example, Coulomb's law involves the products of charges, so the coulomb scale enters twice. In quantum field theory, electric charge can be defined as a numerical fraction of two of nature's fundamental constants: the speed of light (c) and Planck's quantum of action (h). To understand this remarkable claim first see how charge is defined in terms of force. The inverse square law for electric charges shows that the force is proportional to the square of the charge divided by the square of a distance. This means that the dimensions of charge-squared are the same as those of force multiplied by the dimensions of distance-squared. We can summarize this succinctly if we use square brackets to denote 'dimension of', with [L] traditionally denoting that of 'length' (linear distance), so

$$[\text{electric charge}^2] = [\text{Force}] \times [\text{L}]^2$$

Now, Newton's laws of mechanics tell us that a force applied to a body over some distance gives the object kinetic energy. So, the dimensions of force, length, and energy match like this:

$$[\text{Force}] \times [\text{L}] = [\text{Energy}]$$

which implies

$$[\text{electric charge}^2] = [\text{Energy}] \times [\text{L}]$$

If we both multiply and divide by the same quantity, nothing changes. Suppose the quantity is time and we write the accounting of the dimensions like this:

$$[\text{electric charge}^2] = [\text{Energy} \times \text{T}] \times [\text{L/T}]$$

On the right-hand side of this equality, the second term has the dimensions of distance divided by time, like miles per hour: speed. The first term might seem unfamiliar but touches on the foundations of quantum mechanics—Heisenberg's uncertainty principle, which implies that the conservation of energy can be put on hold for a brief amount of time so long as the product of energy and time does not exceed Planck's quantum of action, h. So, if c denotes the speed of light, the dimensions of these are all the same:

$$[\text{electric charge}^2] = [\text{Energy} \times \text{T}] \times [\text{L/T}] = [h] \times [c]$$

QED is the relativistic quantum theory of the electromagnetic field. As such, it combines electric charge, relativity (c), and quantum mechanics (h). It is natural therefore to quantify the squared amount of electric charge as a numerical multiple of $h \times c$. Traditionally a factor of 4π is included but in essence the strength of the electromagnetic force is scaled by the number:

$$(\text{Charge of the electron}^2) \,/\, (h \times c) = \alpha$$

which is known as the fine structure constant, and experimentally is approximately 1/137.

The strength of the electromagnetic force between two doubly ionized atoms, where two electrons have been lost leaving each atom—ion—with a net charge of 2, will be four times that of this basic scale, that between two triply ionized atoms being nine times, and so on.

The strength of the 'strong' force between two quarks will be proportional to the product of their colour charges, analogous to the electromagnetic force being proportional to the electric charges. In turn, the strength of the weak force between quarks or particles like the electron is proportional to their flavour charges. Once the magnitude of this force is known at some distance, it can be compared to the strength of the electromagnetic force whereby the relative strengths of electric and colour charges can be determined.

It is traditional to express these in a similar fashion, thus the resulting definitions of charges for colour and flavour charges lead to analogous numbers, which we could denote by α_c and α_f, respectively. Empirically, when the innards of a proton are studied at a resolution of about 1/100 of a proton's diameter, α_c is roughly 1/10, an order of magnitude stronger than the electromagnetic force. The measure of the quantity α_f is more subtle, because the W and Z bosons have large masses, unlike their massless analogues—photons and gluons—in QED and QCD. In consequence, the weak force does not obey an inverse square law but dies off exponentially fast even over the size of an atomic nucleus. Therefore, it appears to be relatively 'weak' over atomic distances and beyond.

To reveal the fundamental strength of the charge α_f also requires experiments capable of resolving distances 100 times smaller than the size of a proton. Such experiments reveal that the magnitude of α_f of quantum flavourdynamics—QFD—is larger than the α of QED but smaller than the α_c of QCD. This result reveals a tantalizing regularity. Not only do the three forces follow the same basic principle of charges, differing only by the single, dual, and triple nature of those charges for electric, flavour, and colour, respectively, but their relative strengths respect that pattern. The 1, 2, 3 pattern of single electric, dual flavour, and triple colour charges, respectively, is also the order of their respective empirical strengths as expressed by α, α_f, and α_c. Indeed, to recognize this phenomenon it will be worth renaming them α_1, α_2, and α_3, respectively. As a clue to what awaits us, I should add the caveat: this is true at the level of resolution accessible to present experiments or, in other words, at the energies to which we are currently limited.[1]

****

When physicists calculate using QFD, they must keep account of the flavour charges of the various particles involved. The underlying presence of the fundamental two-ness is recorded by a mathematical accounting scheme known as SU2—pronounced ess-you-two. The details of this aren't important other than to note the presence of '2' for the 'two-ness'. The same mathematical structures arise in QCD, which is built on the threefold colour charges. The accounting scheme there is known as SU3. The physical similarities are reflected in the common mathematical formalism underlying their equations. For a theoretical physicist, these mathematical patterns reveal profound harmonies in nature's code and inspire the quest for a grander unifying

[1] For what is predicted to happen at extremely high energies, see chapter 8.

theory. For the experimentalist, the strategy is to test these patterns either to find their limits or to clear the path towards that ultimate theory.

The members of a pair have electric charges that differ by one unit. The marriage of electric and weak (flavour) charges gives what is known as the 'electroweak' interaction. The magnitudes of electric charges of any pair are displaced from the flavour charges of +1/2 and –1/2 by a common number. For the leptons it is –1/2, so that the electron flavour of –1/2 and the additional –1/2 give its charge of –1. Similarly for the neutrino, its flavour +1/2 added to the –1/2 gives its electric charge of 0. A similar shift occurs for the quarks where the displacement factor is now 1/6. Adding 1/6 to the up quark's flavour +1/2 gives the magnitude of its electric charge, +2/3; adding to the –1/2 flavour of the down quark gives the –1/3 for its electric charge. The magnitudes of electric charges being displaced from the flavour charges by a number are what mathematicians call 'U1'. The resulting theory of the electroweak interaction is then based on an 'SU2 X U1' accounting scheme (see Table 2).

The maths appends a subscript *L* on the $SU2_L$ to designate that this empirically only occurs for the *L*eft-handed components of the leptons' and quarks' spins. This reflects the violation of parity, mirror symmetry. This is inserted by hand into the mathematics to recognize the empirical fact of parity violation, but there is as yet no agreed more fundamental theory that requires this to occur. Antiparticles, such as antineutrinos, or the positron and positively charged muon, by contrast all act in the right-handed sense. To find the fundamental reason for this empirical mirror asymmetry is one of the deep challenges facing theoretical physics.

Table 2 Flavour and electric charges for leptons and quarks. The magnitude of electric charge is given by the pair of flavour SU2 charges shifted by addition of the U1 factor. The latter is often referred to as 'hypercharge'.

Flavour	Particle	U1 shift	Electric charge
+1/2	ν		0
		−1/2	
−1/2	e		−1
+1/2	u		+2/3
		+1/6	
−1/2	d		−1/3

The displacement in the magnitude of electric charges relative to the +1/2 and −1/2 of the weak flavour SU2 hints at some further link with QCD, which is described by SU3. Leptons have no colour charge, and their electric charges are displaced by 1/2. The quarks' analogous displacement of 1/6 is one-third of this amount. Is nature teasing us here with a hint of some more profound unity between electric and colour charges? The image is of some process taking place in which a lepton could gain colour charge, there being a one-in-three chance of it being red, or blue, or green. The act of fragmenting into three distinct colour charges involves this 1/2 displacement quantity also being shared three ways, giving 1/6 (or in electric charges the +2/3 and −1/3 of the up and down flavours of quark). Currently this is just numerology, but tantalizing, nonetheless. Particle physicists are seeking a consistent theory where this threefold sharing is a natural result of uniting all three forces, building on their familial mathematical structures: U1, SU2, and SU3.

Up to the bottom

These concepts of electric and flavour charges with their tantalizing triple linkage to colour charge are not mere fancies of theoretical physicists. It is apparent that nature has made use of this mechanism on at least three occasions: as electron and electron-neutrino are twins distinguished by their different flavour charges, so are the muon and the muon-neutrino, as well as the tau and the tau-neutrino. The muon and tau have the same electric charge as the electron, the same spin, the same affinity for a neutrino, and the same lack of interest in the colour forces. Not only are their electric charges identical, but their magnetic moments are too once their different masses are allowed for.

What is good for leptons is good for the quarks. Here too, three sets of flavour pairs have been discovered. The up and down flavour doublet has a much heavier analogue, appropriately known as 'top' and 'bottom'. The top with electric charge +2/3 and the bottom with electric charge −1/3 are again completely analogous to the up and down quarks but for their masses. And their masses are immense. The top quark is about twice as heavy as a Z^0 boson, and is the heaviest fundamental particle yet known.

A third pair of quarks has been discovered whose masses are intermediate between the relatively lightweight up and down and the supermassive top and bottom. Known as 'charm' and 'strange', their whimsical names had been selected before this pattern of the flavours had been identified and the Standard Model emerged (see chapter 8). Like the other pairs, the charm quark has charge +2/3 and the down quark charge −1/3. Each of these six quarks carries any of three colour charges.

These three doublets of leptons, and the corresponding three pairs of quarks, are known as the three 'generations' of fundamental particles. Our familiar world is made of the first generation, the analogous members of the heavier second and third generations being unstable, ultimately decaying into their stable first-generation cousins.

Although members of each generation appear to have the same colour and flavour charges, and identical responses to the electromagnetic, weak, and strong forces, they are more than just heavier versions of one another. We have already remarked on this in the case of the muon and the electron. The implication that the muon contains some special property which we might call muon-flavour, which the electron does not possess, goes beyond what we have so far identified simply as flavour charge. For example, the electron and muon have identical amounts of electric charge which in turn is related to their flavour charge. So, flavour charge, as currently understood, is the same for both particles. Whatever it is that distinguishes muon from electron is something more, its nature yet a mystery. The same can be said for the tau and its neutrino, which again have the identical flavour and electric charges as aforementioned, but in addition possess an intrinsic 'tau-ness'.

Analogous remarks can be made for the quarks. Top, charm, and up quarks each carry some intrinsic label in addition to flavour, electric charge, and colour. This label is carried equally by their siblings, respectively bottom, strange, and down. When the muon was discovered, one physicist famously remarked: 'Who ordered that?' Eight decades later, the question remains.

The Higgs boson

Even if all material particles, both real and those that bubble in and out of existence thanks to quantum uncertainty, could be removed from the vacuum, along with electromagnetic radiation and all sources of gravitational fields too, the vacuum would not be empty. A ubiquitous essence known as the Higgs field would remain. Were the Higgs field also to be removed, the energy of the remaining truly empty void would be higher than when flooded with the Higgs field. This is very much counter-intuitive and the idea that by adding something to nothing you can make it more stable sounds bizarre. Nonetheless that is part of the magic of the idea which goes back to 1964 and was proved dramatically by the discovery of the Higgs boson in 2012.

The Higgs boson's relation to the Higgs field is in some ways analogous to the relation of the photon to the electromagnetic field. Add energy to an electromagnetic field and electromagnetic radiation can occur, which in quantum field theory consists of massless particles: photons. Add enough energy to the Higgs field and Higgs bosons will appear. The most noticeable difference is the amount of energy that is needed to achieve these ends. In the case of photons, it is very simple, as merely striking a match or flicking a light switch can produce millions of them. To produce even a single Higgs boson, however, requires the focus of vast amounts of energy—about 125 GeV—in a small region of space.

Our understanding of the early Universe suggests that in the first moments after the Big Bang, when temperatures were vast and the local energy density also remarkably high, Higgs bosons would have been common. But as the Universe cooled and the ambient temperature dropped, so the Higgs bosons were in effect frozen into the background to form what we now call the Higgs field. Experiments at CERN's Large Hadron Collider have been able to resurrect some slumbering Higgs bosons by smashing beams of protons head-on into one another. The energy in those collisions momentarily heated the vacuum to the point where occasionally a Higgs boson bubbled into existence. The boson itself survives only a fraction of a second before decaying to other particles but the nature of these particles confirms that the Higgs boson has the properties that the theory implied. In summary, these experiments confirm that the Higgs field is the source of mass of the fundamental particles—the quarks and leptons, and also W and Z bosons. Were it not for these elements carrying masses the material Universe as we know it could not exist.

Discovery of the Higgs boson, and by implication the Higgs field, is therefore very significant for understanding nature's scheme, and has been widely described as the capstone completing the Standard Model of particles and forces. There are however many questions remaining. Although the Higgs field is responsible for giving masses to these basic particles, we have no understanding as to why they receive their specific values. And beyond explaining how those masses occur in principle, it effectively says nothing about the nature of charge. When electrically charged leptons interact with the Higgs field, they gain a mass. If this mass is very small, we identify the particle as the electron. If the mass is about 100 MeV, we identify it as the muon. And if the mass is about 2 GeV, we identify it as the tau-lepton.

We do not know why they have those particular masses, but mass alone is not enough to distinguish them. As we have seen, the muon is not simply a heavier version of the electron, and nor, it seems, is the tau. Hopefully as we learn more about the Higgs field and the way it gives particles their masses, we may find clues to answer the question: what distinguishes electron-flavour from muon-flavour and tau-flavour?

8

THE END OF THE MATTER

The nuclear force is strong but of such short range that its effects are feeble at atomic dimensions, let alone in the macroscopic world around us. We would be unaware of this force were we unable to probe distances of nuclear dimensions, the 'femto-universe' of some 10^{-15} metres extent. The so-called weak force also is very short range. It only reveals its full strength when distances of about 10^{-17} metres are probed. The energy required to do this is about 100 GeV, and it is under these conditions of energy, or heat, that we find that the weak force is effectively as strong as the electromagnetic force. It was only when studied from afar, with the poor resolution available at low energy, that it appeared to be feeble, nominally 'weak'.

The perceived weakness of the force is linked to the large mass of its W and Z boson carriers. This link between feebleness and large boson masses raises the question of whether there are other forces transmitted by bosons with masses far bigger than present experiments can access, forces whose effects are limited to exceedingly short distances and beyond our present ability to detect. Some Grand Unified Theories—'GUTs'—predict that there are such forces acting over a distance as small as 10^{-30} metres that can change quarks into leptons and hence cause a proton to decay. They also contain an explanation of why the electric charges of proton and electron are perfectly balanced and of opposite sign. As this would answer the basic conundrum that inspired this journey, we need to understand what GUTs are, and how we might test them.

Everything we have met so far—the quantum theory of the electromagnetic force, quantum electrodynamics, its extension that includes the weak force and the analogous theory of the strong force, quantum chromodynamics, all acting on leptons and quarks that are grouped in pairs—is collectively known as the 'Standard Model'. This is not the last word, however. The Standard Model is essentially a summary of how matter and forces behave at the energies so far explored. We suspect that it approximates a richer theory whose full character is yet to be revealed at yet higher energies.

This does not mean that the Standard Model is in some way 'wrong', just that its reach is limited. Within a restricted range it is unimpeachable. The relation between the Standard Model and a more profound theory of reality that will subsume it, may be analogous to the relationship of Newton's theory of gravity to Einstein's general relativity theory. Newton's theory successfully described all phenomena within its realm of application for hundreds of years. Even today it provides an excellent approximation for most practical needs, such as the timing of tides, sunrise, and sunset, of lunar and solar eclipses, even the motion of the planets and the relative movements of whole galaxies as they interact with one another.

Newton's theory is not the complete description, however, for there are examples which it does not easily handle. These include cases when the relative motions of objects are significant fractions of the speed of light, or where the gravitational force becomes exceedingly strong. An example of the former case, where high-speed motion causes subtle differences between precision measurements and the predictions of Newton's theory, is the motion of the innermost and fastest-moving planet, Mercury. Mercury's orbit does not close; its perihelion—the point in its orbit when it is closest to the Sun—shifts from one orbit to the next. The amount of this precession disagrees very slightly with the prediction of Newton's theory, but general relativity theory successfully explains the minute difference. Newton's theory is not wrong; it emerges from Einstein's theory of general relativity when the effects of the finite speed of light are negligible. In Mercury's case, they are not.

We suspect that a similar feature applies to the Standard Model. The model describes phenomena from energies on the atomic scale at fractions of 1 eV up to 10 TeV—that is 10 trillion eV—the highest energies which terrestrial experiments can currently reach. Throughout that range of energy, spanning more than 13 orders of magnitude, the Standard Model has had total success. However, it has no explanation for the magnitudes of the basic particles' masses nor of other parameters such as the strengths of the forces. These quantities appear to be critical for the emergence of life, but their origins are not understood. Instead, their measured values are input into the equations of the model. Nor does the Standard Model explain why the electric and proton charges counterbalance so perfectly. These puzzles will hopefully be explained someday by a richer, deeper theory of which the Standard Model will emerge as an approximation.

The first steps towards this goal have been the development of GUTs. As Newton's theory is excellent when speeds are small relative to that of light, so the Standard Model is successful when energies are small relative to some 'unification scale'. The precise size of this critical energy is debated by theorists but there is general agreement that it lies below the Planck scale of 10^{19} GeV, the energy at which gravity can no longer be ignored for the interactions of individual particles. Somewhere in the region of 10^{15}–10^{16} GeV seems favoured when the implications of precision measurements from present experiments are extrapolated to extreme energies.

Although the mathematical similarity of QCD, QED, and its extension to include the weak force, quantum flavourdynamics (QFD), hints at some underlying unity among them, the absolute strengths of these forces are very different in practice. So how can any hope emerge that they are ever to be unified? A clue is that we have only measured these strengths over a range of energies that is limited relative to that of the Planck scale. The unification hypothesis is based on the way that the trio of quantum field theories predict these forces behave as energy increases.

Clouds of charge

The simple image of the electromagnetic interaction taking place between an electron and a photon at a point belies a much more profound reality. According to QED, the electron in question is not alone in the void. A vacuum is not empty but seethes with transient particles of matter and antimatter which bubble in and out of existence. Although these will-o'-the-wisps are invisible to our normal senses, they disturb the photon and electron in the moment of their union. QED implies that as you voyage towards that elusive entity of a point of electric charge by increasing the resolution of your measurement, which in effect is the same as increasing the energy of the probe, you do so in the presence of an unseen swarm of ghostly spectators. These include the electromagnetic fields surrounding the electron, plus virtual electrons and positrons, or quarks and antiquarks, or indeed any electrically charged particle and its antiparticle, bubbling in and out of the vacuum.

What a physicist interprets as the electron's charge depends on the resolution of the measurement as determined by the energy of the experimentalist's probe. What we conventionally refer to as the charge on the electron or the proton is the result of a macroscopic measurement, in effect of normal experience. The effects of quantum field theory on this measurement only begin to become apparent when one probes with resolution much smaller than the size of an atomic nucleus. In energy terms, this means experiments with electrons at energies more than 100 GeV. QED theory implies that in such circumstances the effective charge of an electron will have increased in magnitude by about 5% relative to that in a macroscopic measurement. Experiments at CERN, where electrons and positrons collided head-on at such energies, confirmed this. The result of this energy-dependent charge for the individual charged particles, however, implies that as experiments are performed at higher and higher energy, the effective electric charges grow. In turn, this means the electromagnetic forces between charged particles become stronger than what we have been accustomed to in our present restricted low-energy experience.

The same is true for the quarks. Their electrical charges increase with energy, causing the electric charge of the proton that contains them to do likewise. The very mechanism that causes the electron's charge to grow with energy does the same for the quarks, and in perfect synchrony. The result is that the charges of proton and electron always maintain their counterpoise; the combined charges of electron and proton remain at 0 at all energy scales.

The quarks also feel the colour force, which is described by QCD. The dependence of their colour charges on energy is tantalizingly opposite to that of the electrical charges. As energy increases, the effective colour charge decreases. So, what we recognize as a strong interaction at low energies between particles with colour charge becomes enfeebled at higher energies.

The contrasting behaviour is because of a profound difference in the electromagnetic field surrounding electric charge and its colour analogue around coloured quarks. The electromagnetic field has electric charge at its source but does not leak any of that charge itself. In other words, electric charge is the source of photons, but photons are not themselves electrically charged. Contrast the case of colour. Quarks are the source of a colour field which itself carries colour charge. Gluons, QCD's analogue of the photon, carry colour charge.

This causes gluons to interact with one another by colour forces even as they are creating the colour field. A consequence is the difference in the long-range behaviour of the colour field and the electromagnetic field that we noted earlier: an electric field's strength dies away in proportion to the square of distance, whereas the potential energy in a colour field grows with distance. As coloured quarks become separated by some 10^{-15} m, the force between them is so strong that they are entrapped, forming the prisons known as proton or neutron.

The effect of the colour field is to leak colour away from the source into its peripheral surroundings. The higher the energy of a probe, or the more powerful the spatial resolution, so the smaller is the localized colour charge. The result is that at high energies, which can probe this trifling distance, the colour charge becomes enfeebled.

As the powerful colour force has become weaker while the relatively feeble electromagnetic force has become stronger, at what point might these two strengths become the same? This is not possible to answer with certainty because we do not know what other currently unknown particles with very high mass remain to be found. Although these very massive particles may be beyond the reach of any accelerator to produce in experiments, nonetheless they can momentarily fluctuate in and out of the vacuum thanks to quantum uncertainty. In so doing, they could disturb the spread of colour charge or modify the shielding of electric charge and modify the energy dependence of these quantities' rate of change. If we use only the particles that we know exist, and assume that nothing else remains to be found, then the electromagnetic and colour forces are predicted to have the same strength at an energy of around 10^{15} GeV.

This is the first hint that the idea of unification has some meaning: the two forces are described by a common mathematics and have similar strengths at 10^{15} GeV. The unity is obscured at the cold, low energies to which science has so far been restricted.

It was not until the final quarter of the twentieth century that accelerators were capable of colliding particles at energies exceeding 500 GeV and it was only with the advent of these high energies that the changing strengths of the forces began to be measured. We are never likely to be able to collide individual particles at 10^{15} GeV in the laboratory and see the full glory of unity of the forces, but the hot Big Bang model of creation implies this was common back then. Head-on collisions at such energies would have been abundant in that epoch. This has exciting consequences for cosmology as it implies that in the high temperatures of the hot Big Bang there was indeed a unity among these forces which has become hidden as the Universe cooled.

The 1, 2, 3 of unity

So much for QED and QCD, the physical manifestations of quantum field theories built on the mathematical structures U1 and SU3. What does theory imply about the energy dependence of the weak force, governed by SU2? We now know that its historical manifestation as 'weak' is because the W boson—the weak force's analogue of the photon—is very massive, and that experiments of the order of 100 GeV and above are required to reveal its intrinsic strength. Collisions of electrons and positrons in the 1990s at LEP—the large electron-positron collider at CERN—which directly produced its neutral partner, the Z^0, confirmed that once these large energies have been attained, the so-called 'weak' force is indeed comparable to or even stronger than the electromagnetic.

Theory implies that if a unified force at high energies 'freezes' into distinct components at lower energies, with their mathematical structures based on U1, SU2, SU3 in that order, the larger that number, so the stronger the remnant will be. That the relative strengths of electromagnetic, weak, and strong colour forces at moderate energies are so ordered is consistent with them being the frozen remnants of some grand force that applies at very high energies.

Indeed, when the theory is applied and its predictions for the empirically measured strengths of this trio of forces are extrapolated to very high energies, we find that the electromagnetic and weak forces come together at about 10^{16} GeV. This is somewhat larger than the 10^{15} GeV where the electromagnetic matched the strong colour force, but nonetheless in the range 10^{14}–10^{16} GeV, the strengths of all three forces are predicted to be the same to within a factor of two.

They do not appear to be identical at any single energy, however. It is possible that this is an indication that the extrapolation is missing something. Its main assumption was that there are no further massive particles all the way from present energies up to these vast ones. Given the many structures and phenomena that have been revealed in the 12 orders of magnitude of energy spanned during the last century, it would be surprising if there were nothing more from here all the way to the Planck scale of 10^{19} GeV.

Many theorists suspect that there are new families of particles—the 'supersymmetric' partners of the presently known particles—that are potentially to be found at higher energies than those yet accessible. While the theoretical reasons for this are powerful, there is no certain estimate of what their masses are. If they exist, then through quantum uncertainty they will affect the extrapolation of the forces' strengths. If the masses of these particles are some 1000 GeV, and they are then included in the theoretical extrapolation, it turns out that the electromagnetic weak and strong colour forces can indeed be brought into unity at around 10^{16} GeV. If unity is indeed the way of nature, then indirectly this might be a hint that supersymmetric particles are waiting to be found. For our purposes, however, all we can safely say is that while the grand unification concept appears to be very likely, the precise energy at which it is achieved is not well determined. When making calculations that depend on this grand unification, it is conservative to assume that the scale is somewhere in the range of 10^{14}–10^{16} GeV.

In the grand unified world

What does unification imply? The general idea is that the disparate phenomena identified at relatively low energies are fragments of a single circumstance manifesting at very high energy. In the case of the electromagnetic and weak forces, for example, their unity at high energies is masked at room temperature due to the 100 GeV or so of energy needed to free the W boson, the carrier of beta radioactivity. The philosophy of GUTs is rooted in a similar concept, but where the critical energy scale at which the breaking occurs is now 10^{14}–10^{16} GeV. We know that the pattern after the fracturing involves flavours operating in pairs—the SU2 field theory of leptons and of quarks—and colours in triplets—the SU3 of three quark colours which is absent for the leptons. The main question is: how do these separate pieces fuse together above the critical unification energy?

There is no unique way theoretically to merge the various quarks and leptons into grand families above the unification energy. Much mathematical effort has been spent investigating the implications of different groupings. The actions of the forces at low energies reveal a disparity that a unified theory must explain. First, electromagnetism and the neutral part of the weak remnant preserve both the electric charge and colour of the fundamental particles, whereas in beta decay, while the weak force preserves colour, it alters electric charge. The SU3 colour force, meanwhile, acts in the opposite way: it alters colour but preserves electric charge. In our low-energy experience, the combinations of preservation and change are lacking one: there is no known force whose action changes both electric and colour charge. A unified theory must complete this symmetry by containing such a component.

To achieve this there must be analogues of photons, gluons, and W and Z bosons, which carry both electric and colour charge. These consequential agents of GUTs are known as 'leptoquarks'. The name recognizes that the simultaneous alteration of electric and colour charges can transform leptons with no colour and with zero or integer charges into coloured quarks carrying fractional electric charges. This can, of course, happen in reverse, transforming quarks into leptons. This feature can cause the proton, made of quarks, to decay.

That protons will decay due to such a feature appears to be general whatever the mathematical pattern of nature's scheme. The products of proton decay, however, will depend on the specific membership of the grand families of quarks and leptons above the unification scale. The mathematics of group theory encodes the ways the various particles group themselves into families. Any mathematical group will do, so long as it contains as 'sub-groups' the U1, SU2, and SU3 that appear as its separate components at lower energies. The simplest such group is SU5, whose basic family—known mathematically as the 'fundamental representation'—consists of five members, comprising a single flavour of quark in each of three colours and a pair of leptons with no colour.

If you wish, you can find the pattern of particles in SU5 Grand Unified Theory in Figure 10, and accept it as given. However, to understand why it turns out like this is worth a small detour.

A key feature in a quantum field theory described by the mathematical group SU(N), where N is any integer, is that there are $N^2 - 1$ carriers of the force. For example, the trio $(2^2 - 1)$ known as W^+ W^- and Z^0 in the case of SU2, or eight $(3^2 - 1)$ coloured gluons in SU3. In SU5 there will be 24. Twelve of these will be the above set and the photon, the remaining 12 being 'leptoquarks' that carry both colour and electric charge. The mathematical properties of SU5 imply that if electric charge is conserved, the sum of all electric charges in any representation vanishes. So, the electric charges of the fundamental quintet above must add to zero.

This gives a simple equation for the electric charges, Q:

$$Q\ (\text{neutrino}) + Q\ (\text{charged lepton}) + 3Q\ (\text{quark}) = 0$$

So, the quark's charge must be $-1/3$ that of the charged lepton. This is achieved by the members being the down antiquark in each of three colours, an electron, and a neutrino. This link between leptons and quarks has also resolved the conundrum of atomic neutrality, through its linkage of threefold colour and third fractional electric charge. This constraint has come from the leptons and quarks being members of a single family in SU5. It has the unavoidable consequence that emission or absorption of a leptoquark can convert quarks into leptons and trigger proton decay. This is not unique to SU5, but this mathematical group is perhaps the easiest one with which to illustrate the phenomenon.

Two questions probably have come to mind already. First, the family of five doesn't account for all the leptons and quarks: where are the rest of them? Second: what determines the rate of proton decay?

SU5 contains a fundamental family of five but can also contain a group of ten and a singleton. Empirically, the SU2 subgroup that controls the weak interactions at presently accessible energy does not satisfy mirror symmetry. Only left-handed quarks and leptons feel the charge-changing beta radioactivity, whereas the corresponding interaction of antiquarks and antileptons involves only their right-handed components. So, the quintet above, which links electron and neutrino while leaving the three colours of the charge +1/3 antiquark in isolation, refers to the left-handed components. The left-handed components of the up antiquark, the up and down quarks, and the positron give a family of ten (the quarks and antiquarks each coming in any of three colours gives nine in all, plus the uncoloured positron). The left-handed antineutrino is isolated as the singlet.

$$\begin{array}{llll} & 5 & 10 & 1 \\ \text{L:} & [\bar{d}\times 3;\ \bar{e},\ \nu] & [\ (d,u)\times 3,\ \bar{u}\times 3, e^{+}] & [\bar{\nu}] \\ \text{R:} & [d\times 3;\ e^{+},\ \bar{\nu}] & [\ (\bar{d},\bar{u})\times 3,\ u\times 3, e^{-}] & [\nu] \end{array}$$

Figure 10 The quarks and leptons in SU5 Grand Unified Theory. Two quarks in any of three colours and two leptons with no colour gives a total of eight. Inclusion of antiparticles gives a total of 16. These all have spin 1/2 so can occur either right- (R) or left- (L) handed, giving 32 possibilities of fundamental particles and antiparticles in all. In SU5 theory these 32 elements belong to three separate families of left-handed particles and three 'mirror' families of right-handed: a fundamental family of 5 (hence the '5' in the mathematical name), a larger family of 10, and a lone singleton.

The way the fundamental particles and antiparticles organize into families is illustrated in Figure 10. To see why they occur this way, let's check that the total electric charge in each family is indeed zero. We already saw this for the quintet above, and for the singlet, containing an isolated left-handed antineutrino (or in the right-handed analogue, a neutrino), this is obviously the case. For the (left-handed) family of ten, a down and up quark carry net +1/3, and as this pair occurs in any of three colours, this gives a total electric charge of +1 for them. The up antiquark has charge –2/3, three times over, giving net –2. Finally, the positron with its unit positive charge completes the set, the arithmetic +1 – 2 + 1 indeed accumulating to zero.

This distribution of the left-handed components contains a mixture of particles and antiparticles. An analogous distribution occurs for the right-handed set of corresponding antiparticles and particles. The SU5 GUT distributes the particles among the quintet and ten in these unusual combinations, which implies that mirror symmetry is broken for matter, as experimentally observed. So, the empirical violation of mirror symmetry in a Universe made of matter has a natural explanation in SU5: mirror symmetry is only preserved when combined with a switch between particles and antiparticles.[1]

In SU5, the presence of quarks, antiquarks, and leptons in the same family—of ten, for example—hints that what we had regarded as sacred independent conservation rules of quarks and of leptons is a feature that is only manifested at low energy. In general, the only conserved quantities are the *total* number of quarks and leptons, and their electric charges. As the electrically charged W bosons connect different quarks or leptons in

[1] This is not the whole story, see chapter 10.

SU2 theory, so do the leptoquarks of SU5 connect the various members of a family to one another.

For example, a leptoquark can connect the up quark to an up antiquark in the family of ten. Electric charge is conserved, which tells us that in this case the leptoquark must have charge 4/3 to convert the 2/3-charge up quark into an up antiquark with charge –2/3. It also carries colour to account for the colour mismatch between the quark and antiquark. If a down quark absorbs this electrically charged, coloured leptoquark, the quark can change into the other member of the ten family: a positron. When we combine these two steps together, we can see how the proton decays (Figure 11).

How likely or unlikely is this to happen? The SU5 unified theory describes the quantum processes between quarks and leptons that cause the proton to decay, so by using the tools of quantum field theory it is possible to calculate the probability that

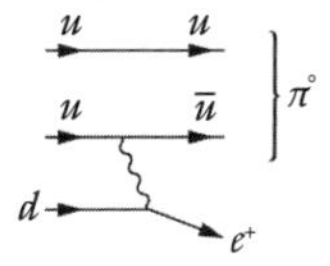

Figure 11 Proton decay in GUTs. A proton can be considered as two up quarks and one down. One of these up quarks turns into the up antiquark by emitting the leptoquark; the leptoquark in turn is absorbed by the down quark, converting it into a positron. What initially was two up quarks and a down quark has become a positron accompanied by an up quark and an up antiquark. This latter combination of quark and antiquark could form an electrically neutral pion, which in turn can decay into two photons. The net result is that a massive proton has converted into a lightweight positron, emitting the excess energy in a flash of light.

the proton will have decayed into a positron and photons after a certain period. To do so we need to know the strength of the leptoquarks' couplings to leptons and quarks.

We know the strength of the quarks' or leptons' couplings to a photon, W and Z bosons, or gluons at low energies, and by extrapolating these to high energies the common strength of their interactions at the unification scale is computed. This unification theory implied the need for leptoquarks, and the very unification relates this common strength to that of the particles' couplings to the leptoquarks too.

If SU5 were not broken, so that it remained the full description even at low energies, the leptoquarks would have been massless, the strength of their interactions very powerful, and protons would have lived for mere fractions of a second. Thankfully, SU5 is broken at around 10^{15} GeV and theoretical models of the dynamics imply that the leptoquarks have masses of this order.

The 100-GeV mass-scale of the W boson limited the effective range of the weak force to about 10^{-17} m. This is much smaller than the size of a proton, but as beta decay involves just a single quark, the overall size of the proton is not critical to beta decay. Proton decay, on the other hand, involves all of its constituents; in other words, the force has to act over the full extent of a proton in order to destroy it. The extreme unification scale of 10^{15} GeV limits the effective range of the proton-destroying force to perhaps 10^{-30} m. As two quarks are involved (Figure 11) they would have to be within that trifling distance of one another. This is very unlikely, which makes proton decay relatively improbable. Proton decay is caused by a force whose range is far too small for the most sensitive instrument to resolve.

Although experiments at present energies are unable to resolve the distances at which such forces act, the effects of such forces might still be detected because quantum mechanics comes to our aid. When we say that a proton is 10^{-15} metres in size, we are making a statement of probability. According to quantum mechanics, about once in 10^{30} years the quarks in a proton will find themselves within 10^{-30} metres of one another and then this force can act on them, causing the proton to decay. Thus, we can study very short distance phenomena if we have great patience and wait for the 'once in a blue moon' configuration to happen.

If protons decay into other particles, it means that all matter in the Universe has a finite lifetime and will eventually decay. There are so many protons in even a small amount of matter—even a dust speck has over a quintillion, 10^{18}—that anything less than exceeding stability would lead to a noticeable erosion of things. Indeed, if their life were less than 10^{17} years, we would be destroyed by radiation from the decaying protons in our own bodies. We would literally feel it in our bones.

A period of 10^{17} years is ten million times longer than the span since the Big Bang, in effect since the start of what we recognize as 'time'. At first sight it may seem paradoxical that our bodies, which live for a mere century, can reveal that protons live so long. However, statistical probability provides the key. The 10^{17} years refers to an average life, technically the half-life, where in a large collection of protons half will have decayed by that time. There are so many protons in our bodies—some 10^{27} or so—that unless protons are intrinsically exceedingly stable, many of them will die young and thereby kill us. It is the combination of rarity—that an individual proton decays—and the vast numbers of them within each of us that gives the conclusion: that we are here at all shows that on average protons must be stable for at least 10^{17} years.

With enough protons we can tip the odds in our favour and need not wait so long. For example, if we have 10^{30} protons, as in a huge container of pure water or in a mass of solid material, there may be one or two protons that decay each year—if these theories and their numerical estimates are correct, of course! The challenge is then one of seeing this event and ensuring that it can be distinguished from other natural background effects. This is what has led physicists to go underground, shielded from cosmic rays, in the hope of finding evidence of proton decay.

9

UNDERGROUND PHYSICS

Until about 10^{-38} seconds after the hot Big Bang, the ambient energy in the Universe was probably as much as 10^{16} GeV. This is some 12 orders of magnitude higher than the energy attainable at the Large Hadron Collider and vastly beyond anything that will be reached by accelerators on Earth in any foreseeable future. If the estimates in the previous chapters of the energy at which a Grand Unified Theory (GUT) applies are any guide, then at that early epoch, a GUT governed the physics of the Universe.

GUTs that unite the electromagnetic and strong forces also unite the fundamental pieces of matter by relating quarks and leptons. This implies, as we saw in chapter 8, that quarks can become leptons and that consequently protons can decay. The prediction of proton decay is one of the few ways that we can test GUTs with present technology.

The actuarial property of quantum physics was the basic idea behind a series of experiments that began in the 1970s. If we take several thousand tonnes of material, instead of the few tens of kilograms in our bodies, and surround this material with sensitive detectors, the giant might feel proton decay in *its* bones.

The key question for any experiment, of course, would be how rapidly—or perhaps more appropriately, how slowly—is the decay expected to happen? This is the biggest challenge for the theorists. All we know for sure is that the GUT which operated early on, in the extremely hot aftermath of the Big Bang, has in our present Universe fractured into two fundamental parts, one governing the electroweak force and the leptons, including the electrically charged electron, the other part of the fracture giving rise to the strong force involving the quarks. Recall, the mathematics that describes the electroweak force is built on the group theory known as SU2 while the mathematics of colour charge is like the electroweak case but built on SU3. The simplest way for this fracture to be recombined is mathematically to have a theory using SU5 where $5 = 2 + 3$ brings us to the world we now know.

When excitement about building viable GUTs emerged in the 1970s, the first calculations assumed the simplest GUT: SU5. The results were exciting because they gave the encouraging prediction that the proton's average lifetime might be as 'brief' as 10^{28} years. For any individual proton this equates to about 10^{18} Universal lifetimes, but thanks to the vast numbers of protons in bulk material, this prediction promised that proton decay might be within the realm of experimental discovery. Uncertainties in the computation, such as the precise energy at which the original GUT was broken, led to a spread of possibilities, the predicted lifetime being in the range 10^{28}–10^{32} years. Nonetheless, this gave hope of seeing a few protons dying young, as over one year between 1 and 1000 protons should die in a kiloton (million kg) of matter. Interest in the possibility of proton decay was also driven by what turned out to be an erroneous experiment that made people think that in a kiloton of material one proton could decay each day, in line with the prediction. If that had been the case, spotting the phenomenon would have been relatively easy. But GUTs' calculations soon established that proton decays were going to be much harder to find.

Other mathematical schemes than basic SU5 were investigated. First was the possibility that SU5 is merged with ideas from 'supersymmetry'.

Supersymmetry theory postulates that for every boson there exists a fermion, and for every fermion there is a boson. There are some compelling theoretical ideas underpinning this, yet there is no direct experimental evidence for such particles. At the very least it is a symmetry that must be severely broken, as the mass of any supersymmetric particles must be much greater than anything yet accessed in experiments. I mention supersymmetry here simply to illustrate how it introduces further uncertainties into predictions of proton decay. The predictions of lifetime cover a similar range to those in theories without supersymmetry, but instead of the proton decaying to a positron and photons, the most likely decay is predicted to be into a neutrino and a strange electrically charged K^+ particle. Another mathematical theory of unification included a different structure, 'SO10', and this gave a much wider range from 10^{30} to 10^{35} years. Some theorists even feared in this scheme the lifetime could be as much as 10^{40} years.

Even with huge amounts of material, it was obvious that protons would decay very rarely, if at all. If one did, however, and you were fortunate enough to detect it, you would have to watch very carefully to be sure that what you saw was not something else faking the real thing. To catch such a faint whisper as proton decay, you must first blanket out background noise such as natural radioactivity and cosmic rays hitting the apparatus, either of which can cause signals that mimic proton decays.

For example, you won't see unambiguous evidence that a proton has decayed, in the sense of there being a gap in some material where previously there was a proton. Instead, you will see evidence of the debris, such as a positron and one or more gamma rays, whose total energies combine to equal that of a proton at rest. Also, conservation of momentum implies that their total momenta add to be the same as that of the proton that disappeared. If this is the result of a free proton decaying—such as one in a hydrogen atom—it will have been effectively at rest because at room temperature a massive proton has negligible motion. If the proton was bound in a nucleus, of oxygen for example, it is unlikely to have been at rest, however. The constituents of a bulky nucleus jiggle due to quantum effects. This random motion will smear out the anticipated momenta of positron and photons when such a proton decays.

Being certain that the signal has come from a proton decay is not so easy, therefore. And there are further problems. For example, it is possible that you were unable to detect all the products. If some energy and momentum are carried off by neutrinos, instead of just positrons and photons, you will be unlikely to detect them, in which case the accounts will not immediately indicate that a decaying proton was responsible. To reduce spurious signals caused by natural radioactivity or by cosmic rays, the material that hosts the protons needs to be extremely pure, any residual radioactivity being recognizable and eliminated as potential signal. As for cosmic rays, by housing the experiment deep underground, the shield overhead of hundreds of metres of solid rock can eliminate all but the most energetic rays, so physicists have gone down mines and under mountains in their search for the signs of a dying Universe.

Experiments begin

The first experiments dedicated to searching for proton decay were built in the 1970s. In India, physicists occupied a 2300-metre-deep mine in the Kolar goldfields, near Bangalore. In Japan, scientists built apparatus in the Kamioka metal mine. In Europe, one team used a cavern off the Frejus tunnel in the French Alps while another took over a garage off the road tunnel 3000 metres below the summit of Mont Blanc. In the United States, detectors were ensconced in the Soudan iron mine in Minnesota, and in a salt mine 600 metres below Lake Erie near Cleveland, Ohio.

In these caverns sat huge swimming pools of water, or monoliths of concrete and steel, containing billions upon billions of protons. The physicists hoped that within these masses a proton might die at some point during the years. And if that key event happened none of the circling bats and insects that live in the caverns would notice, but an electronic device would be triggered and record the occurrence on magnetic tape. Later, in the comfort of their offices, the physicists would examine the data.

The Morton Thiokol salt mine in Ohio was home to the largest of the swimming pool detectors. The device was built by physicists from the Universities of California in Irvine, and Michigan, and the Brookhaven National Laboratory. The IMB detector, as it became known, occupied a huge cavern hewn out of the rock salt by a special mining machine. Two thousand bolts were driven into the rock to prevent the walls from collapsing during the excavation. The Cave formed a cube with sides 20 metres long, as big as a seven-storey apartment block.

Two strong layers of polyethylene lined the walls, in effect forming a huge sack to hold the water. Once in place there came the task of filling it with 8000 tonnes of specially purified water. It takes several minutes to fill a household bath, even with the taps full on. Filling your local swimming pool takes several hours. To fill the IMB tank took two months, and that was only at the second attempt, as the first try burst the polyethylene lining. To prevent this mishap reoccurring, engineers poured concrete behind the lining at the same time as the water level rose inside the bag. The concrete helped resist the pressure from the growing wall of water and stopped it over-stressing the plastic.

Once the tank was full, the physicists and technicians became divers as they installed over 2000 photomultiplier tubes (PMTs) around the faces of the tank. If a proton decayed it would produce charged particles that travel faster than light does in water. This might sound strange because there is a common belief that nothing can travel faster than light. While that is indeed true in a vacuum, light is slowed when it passes through materials such as glass or water. It is then possible that charged particles such as electrons can travel through the water faster than the light can, though still of course slower than nature's ultimate speed limit set by light in a vacuum. When this happens there is a luminous analogue of a sonic boom, and a cone of pale blue light radiates out centred around the flight path. This is known as Cerenkov radiation, named after the Soviet physicist Pavel Cerenkov, whose experiments first led to understanding the phenomenon.

The photo tubes surrounding the IMB tank could detect any sudden bursts of Cerenkov light from a passing charged particle speeding through the water. PMTs act like lightbulbs in reverse. When electric current enters the bulb of a lamp, it gives off light; when light enters a PMT, its energy is converted into an electric current. The resulting electrical pulse can be routed to a computer, which records the event. The record encodes how much light hits each tube and notes the order in which the various tubes are struck. This information enables the computers to work out the directional flow of the light, and in turn to reconstruct the trails of the original culprits. When all this is done the physicists have to assess whether the source was a decaying proton or something more mundane.

The detector at Kamioka was about half the size of IMB, containing some 3500 tonnes of pure water and 1000 PMTs. It was deeper, at about 1 km underground. Even this amount of rock cannot shield apparatus from all cosmic rays, however, as showers of neutrinos get through. Some of these come directly from the Sun or other distant sources, but many are produced by collisions between particles in the cosmic rays and atoms in the upper atmosphere. The Earth is transparent to most of them; neutrinos from the Sun, for example, shine down on our heads by day and up through our beds by night, undimmed. These neutrinos are so numerous that occasionally one interacts with an atom of hydrogen or oxygen in the water, giving a signal that can mimic spontaneous proton decay.

Abdus Salam, an enthusiastic Pakistani theorist working at Imperial College, London, believed strongly that protons decay. He wrote a paper suggesting that this unwanted background of atmospheric neutrinos could be eliminated if an experiment was done on the Moon. One leading neutrino experimentalist who was no respecter of theorists reviewed Salam's paper and commented that if Salam or other theorists wanted to go to the Moon, then let it be, 'the more the merrier'.

Gradually it became clear that the phenomenon of proton decay would be so rare as to be invisible, even if it happened at all. This made these huge detectors deep underground useless for their original purpose. Instead, the teams of experimentalists started to study the neutrinos that hitherto had been the unwanted background.

What happened proved the adage that in a crisis there lurks opportunity. Although the detectors had been designed to look for signs of decaying protons, they proved to be superb means of detecting neutrinos from the stars. The Japanese experiment in the Kamioka mine happened to be especially fortunate. The original experiment was named Kamiokande—'Kamioka Nucleon Decay'—encapsulating its location and its goal.[1] In 1985 its detectors were reprogrammed to be more sensitive to neutrinos and by the end of 1986 Kamiokande was ready for its new task, its name reinterpreted as Kamioka Neutrino Detector. Within months it had made a remarkable discovery, when for a few seconds around 07:30GMT on 23 February 1987, a sudden burst of neutrinos passed through the Earth. We are bathed in a flux of solar neutrinos all the time, but that momentary burst was the blast from a dying star located 170,000 light years away in the Large Magellanic Cloud. The 'LMC', as it is known, is a satellite galaxy of our own Milky Way and is visible in the southern skies. The neutrinos had not only crossed that vast distance of space and time but also passed through the southern hemisphere of the Earth before being detected in Japan. The gravitational collapse

[1] 'Nucleon' is a generic word for the nuclear constituents, proton or neutron.

of a supernova produces a brilliant flash of light which briefly outshines an entire galaxy. Powerful though this is, electromagnetic radiation adds up to less than 1% of the whole; the bulk of the energy radiated by supernovae is carried away by neutrinos.

That these detectors managed to record such a unique event is remarkable in many ways. First, understand that the original explosion occurred not just far away in space, but 170,000 years ago. It had taken but seconds for these neutrinos to diffuse out from the star, which had a density like that of a huge atomic nucleus. That brief blast created a spherical shell of neutrinos, moving almost at the speed of light, whose thickness was a few light-seconds—about ten times the distance from the Earth to the Moon. Travelling more than 10 million miles every minute, it spread out across intergalactic space. Ahead lay the Milky Way, in an arm of which, on the small planet Earth, human life had advanced to the Stone Age. The shell of radiation would travel on for over 165,000 years, by which time people around the Mediterranean were beginning to be aware of the heavens and were inventing science. The wave from the collapsed supernova continued approaching the Earth through the southern heavens. It was 31 light years away when scientists proved that neutrinos exist, and it was less than a light year away when scientists in America and Japan completed building huge tanks of water underground designed to look for signs of proton decay.

It was by chance that detectors designed to capture evidence of decaying protons were ready when the blast from the Large Magellanic Cloud swept through.

The supernova explosion had emitted 10^{59} neutrinos. After their journey of 170,000 years, some of them passed through the Earth and carried on into the northern sky. One thousand trillion neutrinos must have passed through IMB and a similar number through Kamiokande and other experiments. Only IMB and Kamiokande were at that time able to record their passage, however. Even so, most of the neutrinos passed right through the Earth as if it were nothing more than empty space. Of the hordes, just 8 neutrinos were detected colliding with matter in IMB, and 11 in Kamiokande. These handfuls nonetheless were sufficient to confirm astrophysicists' theories of supernovae and heralded the maturity of a new science: neutrino astronomy.

****

By 1990 the score from the detectors was: confirmed supernova, 1; evidence of proton decay, 0. Following the proof that Kamiokande can detect cosmic neutrinos, in the subsequent decades its operation has primarily been geared towards studying them. These ghostly particles were originally thought to be massless, but it's now known that they do have mass, albeit very small. Neutrinos are the least understood of the presently known particles, which is a reason why Kamiokande has focused attention on them. Adding to their interest is the possibility that neutrinos may have been instrumental in breaking the symmetry between matter and antimatter, enabling the asymmetry that is now apparent in our matter-dominated Universe.

In 1995, Kamiokande was closed for reconstruction and enlargement. By this stage the absence of a positive proton signal put its lifetime to decay into a positron and photons at longer than 2.6×10^{32} years. By 1996, Kamiokande was ready for more work with a cylindrical tank of 39 metres diameter and 42 metres high, containing 50 kilotons of pure water and surrounded by more photomultiplier tubes than before. The detectors on its surface now covered an entire acre. The electronics however were specially tuned to get the maximum information about neutrinos—low-energy ones from the Sun and higher-energy ones produced by cosmic ray collisions in the atmosphere. Its primary task as a seeker of proton decay was effectively ended, even though physicists continued to monitor for any signs of the phenomenon. The detector was renamed Super Kamiokande—SuperK for short.

SuperK ran stably for 20 years. After a decade with no proton decay candidates, its volume was increased by 25%. This was done by reducing the space between the cylinder and the cavern walls by half, from two metres to one. By 2018, with still no proton decays, the lifetime of the proton against decay into a variety of potential products was elevated to be greater than 10^{34} years. This is now very near predictions of some theoretical models which imply proton decay should occur with a lifetime of about 10^{35} years.

If indeed protons are unstable with a lifetime of this size, how can we improve the experiment by a factor of ten? It is unrealistic to find the evidence by continuing the experiment ten times longer than so far, in other words, for 200 years! The solution has been to design a larger detector. Where Kamiokande's water tank was 3 kilotons, and SuperK's was 50 kilotons, the decision was taken to build HyperKamiokande—'HyperK'—with a tank of 260 kilotons.

This is being constructed in a separate cavern, about 650 metres below ground. All being well, it should be completed and ready for physics experiments in 2026. HyperK will consist of a cylindrical tank, 60 metres high and 74 metres in diameter, its 260,000 tonnes of ultrapure water filling a volume 75 times bigger than that in the original Kamiokande experiment.

Technology has moved on in the four decades since Kamiokande was designed. This will enable the phototubes to be much more efficient than previously. All in all, HyperK's ability to detect trails from decaying protons and to filter them from background noise will be hundreds of times better than ever before. While its guaranteed research diet will consist of studying neutrinos, it will also provide the most sensitive measurements yet on the stability of the proton. In ten years, it should be able to push limits on proton decay to greater than 10^{35} years, or hopefully see proof of the phenomenon.

10

MYSTERIES

All along in this mystery I have assumed that atoms are indeed neutral and that the charges of electron and proton do perfectly balance. Although we have identified a broad rationale in terms of constituents, families, and patterns, how sure are we that these charges are indeed identical, and what consequences ensue if they are different?

Back in 1924, Albert Einstein inadvertently inspired one of the earliest questionings of this charge equality. He remarked that the magnetic fields of both the Sun and Earth have their orientation and sense closely aligned with their axes of rotation and he proposed that this was the result of circling 'mass currents' generating electromagnetic fields. His motivation was to build a theory where electromagnetism and gravity are the result of a common underlying field. His attempts eventually proved fruitless, but when the mathematics of his theory were applied to these magnetic fields, the formulae turned out to be equivalent to those in the standard theory of electromagnetism but where the absolute magnitudes of electron and proton charges differ.[1]

Einstein's original proposal and the possible charge asymmetry are deeply physically connected. They can be theoretically merged if the charge imbalance, Q, expressed by the formula

$$Q = (Q_e + Q_p)/Q_p$$

is of the order of the ratio of the proton's mass to the Planck mass scale—the energy at which quantum effects can no longer be ignored in general relativity. The proton energy at rest—its 'mc^2'—is about 1 GeV, whereas the Planck scale is some 10^{19} GeV. If Einstein's conjecture were correct, the charge mismatch Q would be some 10^{-19}, that is 1 part in 10^{19}, which is like the diameter of an atom compared to that of the Sun.

It might seem incredible that one could measure a value for Q to this level of precision, but the vast size of Avogadro's number—6×10^{23}—comes to our aid. This is the number of molecules in a mole (molecular weight in grams) of a substance, so testing the electric neutrality of a container of gas, for example, could translate to this level of accuracy at the level of individual molecules. Another idea has been to exploit the cumulative gravitational attraction between large numbers of atoms and compare it to any residual repulsive electrical forces that would arise from a charge imbalance. This idea has had an interesting history in cosmology.

In the 1960s, before discovery of the microwave background radiation, there was much debate about which of the Big Bang

[1] The history of Einstein's idea is fascinating but goes far beyond the story here. It is described in some detail by C. S. Unnikrishnan and G. T. Gillies, *Metrologia*, 41, S125, 2004.

or Steady State models of the Universe was correct. One of the favoured pieces of support for a moment of creation, some 13.8 billion years ago, is the observed expansion of the Universe. The cosmologist Herman Bondi, who was an advocate of the Steady State theory, proposed that the expansion of the Universe might have nothing to do with a Big Bang but instead could arise if the electric charges of electron and proton do not perfectly balance.

His argument was brilliantly simple. Newton's law of gravitational attraction is an inverse square law, as is that between two electric charges. The force of electrical repulsion between two protons relative to their gravitational attraction is therefore a dimensionless number, whose value is of the order of 10^{36}. The strength of the gravitational force is proportional to the product of the masses, while an electrical force is proportional to the product of the charges. Therefore, any electrical repulsion arising from this mismatch is in proportion to 10^{18}—the square root of 10^{36}.

Bondi duly pointed out that if Q empirically were just a factor of ten bigger than Einstein's value, namely 1 part in 10^{18}, electromagnetic repulsion would dominate gravitational attraction among hydrogen atoms on a galactic scale. Electrical imbalance could therefore explain the expansion of the Universe within Newtonian mechanics. He even published a paper on this in the *Proceedings of the Royal Society* (volume A252, page 313). Today, the hot Big Bang model has been established and there is no need to invoke electrical repulsion, so we can now reinterpret Bondi's calculation as giving a cosmological bound that Q must be smaller than 1 part in 10^{18} for hydrogen.

The afterglow of the Big Bang—the cosmic microwave background radiation (CMBR)—has been studied by experiments mounted on satellites. It is remarkably uniform, with fluctuations in temperature a mere 1 part in 100,000. Any charge imbalance between protons and electrons would have left an imprint in the form of vortices in the CMBR, but none have been seen. Translating the absence of such eddies into a limit on the possible size of Q depends on further assumptions. Nonetheless, the resultant estimated limit can be impressive. For example, if any excess charge is distributed uniformly, Q must be smaller than 10^{-29}. However, there is much we do not understand about the material content of the cosmos, notably the roles of so-called dark matter and dark energy. If these are mildly electrically charged, the above result could be nullified.

While measurement of matter's electrical neutrality may have significant implications for cosmology, the most reliable limits for individual atoms come from laboratory experiments.

Einstein originally aired his hypothesis at a meeting of the Swiss physical society, in Lausanne in 1924. At the following meeting, in 1925, the Swiss-born physicist Auguste Piccard reported how he and an assistant, a Dr Kessler, working at the Free University of Brussels, had been inspired by Einstein's conjecture and made a series of measurements to test it. Piccard later became famous for his ascents in air balloons to 18,000 metres (60,000 feet) to investigate the stratosphere, and in the 1940s for his deep-sea dives of more than 2000 metres in bathyscapes. But his first significant contribution to science came with his investigation of electrical balance in 1925.

Piccard and Kessler used a hollow spherical iron container—a 'bombe'—filled with 27 litres of CO_2 at 8 times atmospheric pressure. The 'bombe' was charged to several thousand volts and completely deionized, so it contained the same number of electrons and protons. They housed it inside a quartz tube, opened a tap on the bombe, and let the gas escape. As the vapour leaked out of the container, they monitored the bombe's electrostatic field. The result was that the field didn't change, which implied that no electric charge was escaping. They made a few trials and quickly established Q to be less than 10^{-16} and then 10^{-18}. As their technique improved, their limit became stronger, and at the 1925 meeting in Geneva they announced Q to be less than 5×10^{-21}.[2]

As this is more stringent than Bondi's conjecture, which he made 30 years later, it seems that their article, written in French and published in a rather parochial place of record, escaped wider attention. Further reasons for it being ignored might be because there is no fundamental basis for Einstein's relation anyway. The Planck scale is a natural physical measure, but why should the mass of the proton rather than of the electron, say, be used in the numerator? Were the latter chosen, Einstein's ratio would be reduced to the order of 10^{-22} which is not eliminated by their experiment. That said, there is no reason why Einstein's ratio should have anything fundamental to do with this at all. It has been a convenient numerical standard against which experiments might attempt to set a limit.

The strongest limit currently known to me is already 50 years old. Back in 1973, H. Frederick Dylla and John G. King attempted to detect sound waves in a gas excited by an alternating electric field in a spherical resonator.[3] If atoms have a small charge asymmetry, the frequency of the sound wave will be the same as that of the alternating electric field; for a gas of neutral atoms, however, the frequency will be twice this amount. Their measurements of frequency and of any modulations in the wave

[2] A. Piccard and E. Kessler, *Arch Sci Physiques Naturelles*, 7, 340, 1925.

[3] H. F. Dylla and J. G. King, *Phys Rev*, A7, 1224, 1973.

proved to be a very sensitive way of limiting the presence of a charge asymmetry. Their result improved on Piccard and Kessler by a factor of five, placing Q at less than 10^{-21}.

Neutron, neutrino: neutral?

What about fundamental particles that are believed to be uncharged, such as the neutron? A non-zero charge for the neutron can be sought by passing beams of them through an electrical field. The amount of electric force is proportional to the neutron's electric charge.

This was first done back in 1981. Nuclear reactors produce neutrons, which can be collimated into a beam. The neutrons were moving at about 200 metres per second and travelled some 9 metres through a uniform electric field of strength 6 million volts per metre. The sensitivity of the experiment was limited by the ability to measure any deflection of the beam. The experimenters were able to show this was less than 10 nanometres, which translated into the neutron charge, Q, being zero to an accuracy of about 2 parts in 10^{20}.

In 1987, improved experiments were done, where the speed of the neutrons ranged between 150 and 400 metres per second. The results confirmed that the neutron charge Q is less than 10^{-21}, a similar precision to that on the neutrality of atoms.

Neutrinos are much harder to study directly. For them, the neutrality test uses astrophysical arguments. If neutrinos are electrically charged, photons can convert to neutrino–antineutrino pairs. This would affect the rate of stellar evolution. We know that our Sun has survived for over 5×10^9 years, which places an upper limit on the rate of energy loss by a possible neutrino–antineutrino pair production. Consistency requires the neutrino's charge Q to be less than 10^{-13}.

It's possible to place stronger limits if a piece of standard theory is considered. Electromagnetic interactions are described by field theory which satisfies the constraints of relativity and of quantum mechanics. For these constraints to be satisfied, electric charge must be conserved. This implies that in the beta decay of the neutron into proton, electron, and antineutrino, the charges of proton and electron must equal the sum of the neutron and neutrino charges:

$$Q_p + Q_e = Q_{nu} + Q_n$$

If one accepts charge conservation, this equation provides a better empirical limit on the neutrino charge than the astrophysical argument. The left-hand side is known to be at most 10^{-21}, and Q_n too is empirically of this order, which implies a similar constraint for Q_{nu}. However, there is nothing that requires each side of the relation to equal 0 separately unless leptons and quarks are related by some grand unified theory.

Quantum electrodynamics was based on a U1 theory where the charge is expressed as an arbitrary integer multiple of a basic quantum. The weak interaction is based on an SU2 theory where there is a duality to the basic measure of charge, conventionally these being carried in a positive or negative amount by two sibling particles, such as the up and down quark, or the neutrino and electron. In electroweak theory, where the electromagnetic and weak interactions are treated together, electric charges are not fixed. The quarks and the electron could have arbitrary charges. Recall from chapter 7 that electric charge is now the result of the SU2 charges being uniformly shifted by an amount of U1 'hypercharge'. This hypercharge is completely unconstrained, at least within current understanding. Empirically, the hypercharge of a lepton pair is three times that of the quarks and of opposite sign, which is what leads to the 2/3 and –1/3 electric charges of the quarks, relative to the –1 and 0 of electron and neutrino. This is an empirical fact; there is as yet no explanation for these specific values outside GUTs.

While the overall magnitudes of electric charge are not theoretically constrained, the equality arising from beta decay, namely that the difference of electric charge between up and down quark, or between proton and neutron, is the same as that between electron and neutrino, follows immediately from the SU2 feature. Each of these pairs has its two members carrying the fundamental positive and negative amounts of SU2 charge, and it is this difference that is carried by the W^+ or W^- as they transmit the force.

In GUTs, where the mix includes the triplet of colour charges—the SU3 analogue of the flavour SU2—constraints emerge for the total electric charge. The simplest example of unification is to bring these together in a theory built on SU5. This was illustrated in chapter 8, but there are other mathematical structures that accommodate the known particles, and we have as yet no intuition as to which—if indeed any—of these nature uses. The one general feature, however, is that a GUT marries leptons and quarks together and this requires that quarks, the constituents of protons, can transform into leptons, causing the proton to decay.

In all GUTs the beta decay constraint on the difference of charges for leptons or for quarks must apply. However, their consequences for the neutrality of atoms can differ from SU5, and in intriguing ways. A mathematical scheme known as SO(10) is one such.

The SO(10) scheme is perhaps the simplest one beyond the SU5 example. The constraint

$$Q_p + Q_e = Q_{nu} + Q_n = 0$$

is true, but SU5 was very restrictive in requiring that Q_{nu} and Q_n are each independently zero. In SO(10) and more generally, there remains the possibility that while charges of neutron and neutrino are opposite they are not necessarily zero. If the neutrino has a small charge, and the neutron has a corresponding charge but of opposite sign, a hydrogen atom comprised of an electron and a single proton will be neutral, but atoms of deuterium, whose nucleus contains one proton and one neutron—and of other elements, whose nuclei contain many neutrons—are not.

This shows how careful one must be before drawing conclusions from experiments. Measurements of hydrogen gas escaping in an electric field may well show its charge to be consistent with zero, but one must not assume that this implies the same result could hold for gases made of other elements.

A gutsy worldview

In conclusion, let's assume the present data are correct: neutron, neutrino, and the hydrogen atom each have precisely zero overall electric charge. Also, I include in the list of unexplained phenomena that whereas mirror symmetry—parity—is a symmetry of electromagnetic and colour interactions, it is empirically broken for the weak interaction. Finally, we should confront the largest asymmetry of all: the Universe appears to be made of matter to the exclusion of antimatter, so is there some difference between the properties of quarks or leptons on the one hand, and their 'charge mirrors' the antiquarks and antileptons on the other? What insights might a GUT offer to these mysteries? But before starting this exercise, how out of a host of theoretical possibilities can we select a particular GUT to use, if only as a pedagogic foundation?

In the search for a possible GUT as a tool for this exercise, I shall appeal to history. A summary of the developments in quantum field theory since the middle of the twentieth century is that nature seems to choose the simplest solution. The story began in 1947 with the discovery of QED. The essential basis of QED was that it is a relativistic quantum field theory which, in mathematical jargon, has a U1 structure. Two decades of false trails eventually led to an explanation of the weak and strong interactions. Several fantastic ideas fell by the wayside as it became clear that these two forces are the SU2 and SU3 generalizations of the basic U1 idea that had long been understood in QED.

In the case of the weak force this feature was obscured by the fact that its carriers—the W and Z bosons—are massive. This is now understood to arise due to the presence of the Higgs field. As for the latter, experimentally this too appears to follow the simplest mathematical scheme, namely that which Peter Higgs and others constructed when the idea first emerged, back in 1964.

This simplicity is like Occam's razor being applied to particle theory: assume the most economical path is followed, until data decree otherwise. Within the philosophy of Occam's razor, I will use the simplest candidate for a GUT, SU5, as a basis for confronting our present state of ignorance and seeing what insights it offers about the menu of known particles and their interactions. The failure to have observed proton decay so far could be because it is only at the border of observation and because present estimates of the scale of GUTs within SU5 are misjudged.

If, however, we are someday forced empirically to follow a more complicated path, this would add a further question to the list: why did nature not choose the simplest option?

An obvious example is already on our list: why does mirror symmetry fail for the weak interaction? As we saw in chapter 8, an SU5 GUT can explain this, and in so doing gives a novel perspective on the relations among the fundamental fermions of the Standard Model. The breaking of SU5 into SU2, SU3, and U1 shows that the fundamental symmetry is between left-handed particles and right-handed antiparticles (or between right-handed particles and left-handed antiparticles). If, as is traditional, we refer to mirror parity by the symbol P, and the particle–antiparticle switch—or 'charge conjugation'—by C, the symmetry that acts is CP. C and P individually do not apply.

That we got diverted into perceiving the failure of parity, P, as a profound problem is because we mistakenly thought we had a total picture when, in reality, we are confined to but one half of two C-conjugate worlds. Our matter is the result of electron and neutrino, up and down quarks; the positron, antineutrino, and antiquarks have no lasting presence here. That our lopsided 'C' perspective should be 'P' asymmetric too ought to be no surprise.

The real puzzle is that this P asymmetry is restricted to the weak interaction, whereas electromagnetism and the strong forces are mirror symmetric. The way particles and antiparticles group together in SU5, however, shows that the breaking to U1, SU2, and SU3 leads to P violation being directly manifested by the doublets of electron and neutrino or of up and down quarks. Another way of getting intuition about this phenomenon is to focus on the singleton that occurs in SU5. This must be a colourless particle with no electric charge, namely a neutrino. Being a singleton, it can have no link to an electron via the weak (or indeed any) force. This condemns one handedness of neutrino to sterile isolation (conventionally, the right-handed neutrino), leaving the other (left-handed) to link to the electron.

That the violation of P in our C-lopsided world need not be surprising is not original with me. This viewpoint was quickly adopted by the physics community following the original discovery of P-violation in 1956, and CP as the true symmetry became the new paradigm. This vision held until 1964 when even this symmetry was found to fail in the properties of the strange particles known as K mesons. There is nothing in the SU5 pattern that enables electron, neutrino, or up and down quarks to behave differently to their corresponding antiparticles, viewed in a mirror, so CP violation is to be explained.

Since then, CP violation has been found also with charm and to a larger extent with bottom flavours of quark. In all cases, flavours in different 'generations' are involved. Can even this be understood within the SU5 GUT?

The distribution of fundamental fermions among the families of ten, five, and one so far has always focused on the members of the first generation—electron, neutrino, up and down. SU5 applies equally to the second generation—muon, muon-neutrino, charm and strange—or to the third—tau, tau-neutrino, top and bottom. In isolation each of these has CP symmetry locked in, like we saw for the first generation. There remains, however, the question of how one generation relates to another. The mathematics of complex numbers reveals a tantalizing loophole.

The wavelike descriptions of fermions allow those in the second generation to have a complex phase relative to those of the first, all else remaining the same. The antiparticles will be shifted by the same phase but opposite in sign. For example, if the phase is compared to the positions around the face of an analogue clock, then if the first generation is arbitrarily chosen to be at 12 o'clock, and the phases of the particles in the second generation are at 2 o'clock, their antiparticles will have their phases set at 10 o'clock. Similarly, were the third-generation particles' phases set at 3 o'clock, their antiparticles would be at 9 o'clock. These unbalanced phases enable CP violation to occur when particles of different generations are involved. For example, a strange meson might contain an up quark bound to a strange antiquark. The anti-version of that meson contains an up antiquark and a strange quark. The phase mismatches cause the two to have different properties when, for example, they decay.

Mathematically SU5 can accommodate CP violation, though empirically the magnitude of the phenomenon seems to go beyond this simple example. Experiments are investigating whether the three generations of leptons, in particular their enigmatic neutrinos, might be the key to CP violation thanks to this loophole. Time will tell.

Whether or not this explains CP violation, there remains the enigma of why nature has copied the first generation to make three in all. (Indirect arguments suggest that there are not more than three such generations.) That each generation has some unique label—some further 'charge' perhaps—seems clear, given the failure of a fermion in the second or third generation to convert into its first-generation 'ground floor' counterpart by emitting electromagnetic radiation. We cannot rule out the possibility that a super-weak force, acting over truly trifling distances, is governed by some novel charge that gives each generation a distinct character.

To end, let's return to the mystery that inspired our quest: the neutrality of matter.

Theorists have examined the fine print of GUTs carefully. The conclusion seems to be that the charge of electron and proton must necessarily counterbalance perfectly. The present limits of the order of 10^{-20} are tantalizing and it is by no means ruled out that an improvement could bring unexpected results. For now, at least, all experimental data are consistent with the fundamental conundrum. On the one hand, the only understanding of the hydrogen atom's vanishing charge is that the electron and the proton's constituent quarks are related by some yet to be understood grand unified theory. On the other, the familial connection between leptons and quarks then leads unavoidably to the conclusion that protons can decay. Perhaps the data from Hyper-Kamiokande will find the first evidence of this decay, but if not, this profound paradox will remain unresolved.

Index

For the benefit of digital users, indexed terms that span two pages (e.g., 52–53) may, on occasion, appear on only one of those pages.

Note: Tables, figures, and notes are indicated by an italic t, f, *and* n *following the page number.*

F

G